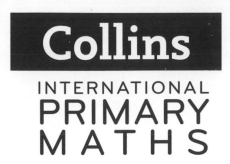

Collins
INTERNATIONAL PRIMARY MATHS

Workbook 6

William Collins' dream of knowledge for all began with the publication of his first book in 1819. A self-educated mill worker, he not only enriched millions of lives, but also founded a flourishing publishing house. Today, staying true to this spirit, Collins books are packed with inspiration, innovation and practical expertise. They place you at the centre of a world of possibility and give you exactly what you need to explore it.

Collins. Freedom to teach.
Published by Collins
An imprint of HarperCollins*Publishers*
The News Building
1 London Bridge Street
London
SE1 9GF

HarperCollins Publishers
Macken House, 39/40 Mayor Street Upper,
Dublin 1, D01 C9W8, Ireland

Browse the complete Collins catalogue at
www.collins.co.uk

ISBN 978-0-00-836950-7

British Library Cataloguing-in-Publication Data
A catalogue record for this publication is available from the British Library.

Author: Paul Hodge
Series editor: Peter Clarke
Publisher: Elaine Higgleton
Product developer: Holly Woolnough
Project manager: Mike Harman (Life Lines Editorial Services)
Development editor: Tanya Solomons
Copyeditor: Catherine Dakin
Proofreader: Catherine Dakin
Cover designer: Gordon MacGilp
Cover illustrator: Ann Paganuzzi
Typesetter: Ken Vail Graphic Design, Ltd
Illustrators: Ann Paganuzzi, Ken Vail Graphic Design and QBS Learning
Production controller: Lyndsey Rogers
Printed in India by Multivista Global Pvt.Ltd.

With thanks to the following teachers and schools for reviewing materials in development: Antara Banerjee, Calcutta International School; Hawar International School; Melissa Brobst, International School of Budapest; Rafaella Alexandrou, Pascal Primary Lefkosia; Maria Biglikoudi, Georgia Keravnou, Sotiria Leonidou and Niki Tzorzis, Pascal Primary School Lemessos; Taman Rama Intercultural School, Bali.

The publishers gratefully acknowledge the permission granted to reproduce the copyright material in this book. Every effort has been made to trace copyright holders and to obtain their permission for the use of copyright material. The publishers will gladly receive any information enabling them to rectify any error or omission at the first opportunity.

Photo acknowledgements
Every effort has been made to trace copyright holders. Any omission will be rectified at the first opportunity.
p18tl OrangeVector/Shutterstock; p18tr Rvector/Shutterstock; p18bl Victor Z/Shutterstock; p18bc RaZZeRs/Shutterstock; p18br Rvector/Shutterstock; p20 Wonderful Future World/Shutterstock; p47 Anatolin/Shutterstock; p135c Jan Martin Will/Shutterstock; p137 Mhatzapa/Shutterstock; p153 Kurilenko Katya/Shutterstock.

Contents

How to use this book

This book is used during the middle part of a lesson when it is time for you to practise the mathematical ideas you have just been taught.

- An **objective** explains what you should know, or be able to do, by the end of the lesson.

You will need
- Lists the resources you need to use to answer some of the questions.

There are two pages of practice questions for each lesson, with three different types of questions:

 Some question numbers are written on a **circle**. These questions may be **easier**. They may also practise mathematical ideas you have learned before. These questions will help you answer the rest of the questions on the two pages.

3 Some question numbers are written on a **triangle**. These questions provide **practice** on mathematical ideas you have just been taught. They help you to understand the ideas better.

5 Some question numbers are written on a **square**. These questions are slightly more **challenging**. They make you think more deeply about the mathematical ideas.

You won't always have to answer all the questions on the two pages. Your teacher will tell you which questions to answer.

Questions with a star beside them require you to Think and Work Mathematically (TWM). You might want to use the TWM Star at the back of the Student's Book to help you.

Date: _____

At the bottom of the second page there is room to write the date you completed the work on these pages. If it took you longer than 1 day, write all of the dates you worked on these pages.

Self-assessment

Once you've answered the questions on the pages, think carefully about how easy or hard you find the ideas. Draw a ring around the face that describes you best.

 I can do this.

 I'm getting there.

☹ I need some help.

Number

Lesson 1: **Counting on and back in fractions and decimals**

- Count on and back in decimal steps and fractions

1 Count on or back in the steps given.

a Count on in steps of 0·2.

2·5, ☐, ☐, ☐, 3·3, ☐, ☐, 3·9, ☐, ☐

b Count back in steps of 0·01.

5·22, ☐, ☐, ☐, 5·18, ☐, ☐,

5·15, ☐, ☐

c Count on in steps of $\frac{1}{3}$.

$6\frac{1}{3}$, ☐, ☐, ☐, $7\frac{2}{3}$, ☐, ☐, $8\frac{2}{3}$, ☐, ☐

d Count back in steps of $\frac{1}{2}$.

$8\frac{1}{2}$, ☐, ☐, ☐, $6\frac{1}{2}$, ☐, ☐, 5, ☐, ☐

2 Count on or back in the steps given.

a Count on in steps of 0·4.

6·7, ☐, ☐, ☐, 8·3, ☐, ☐, 9·5, ☐, ☐

b Count back in steps of 0·03.

4·55, ☐, ☐, ☐, 4·43, ☐, ☐,

4·34, ☐, ☐

c Count on in steps of 0·05.

2·114, ☐, ☐, ☐, 2·314, ☐,

☐, 2·464, ☐, ☐

d Count back in steps of $\frac{2}{3}$.

$7\frac{1}{3}$, ▢ , ▢ , ▢ , $4\frac{2}{3}$, ▢ , ▢ , $2\frac{2}{3}$, ▢ , ▢

e Count on in steps of $\frac{3}{5}$.

$1\frac{1}{5}$, ▢ , ▢ , ▢ , $3\frac{3}{5}$, ▢ , ▢ , $5\frac{2}{5}$, ▢ , ▢

3 Draw a ring around the numbers that are part of the counting sequence.

a Count forwards in steps of 0·04 from 2·11.

2·15 2·19 2·22 2·27 2·3 2·35 2·39

2·43 2·48 2·51 2·54 2·59 2·63 2·68

b Count back in steps of 0·03 from 6·87.

6·85 6·83 6·82 6·8 6·78 6·63 6·6

6·55 6·52 6·49 6·39 6·35 6·33 6·32

c Count forwards in $\frac{3}{8}$s from $2\frac{6}{8}$.

$3\frac{1}{8}$ $3\frac{4}{8}$ 4 $4\frac{2}{8}$ $4\frac{5}{8}$ $5\frac{1}{8}$ $5\frac{3}{8}$ $5\frac{6}{8}$ $6\frac{1}{8}$ $6\frac{5}{8}$ 7 $7\frac{3}{8}$ $7\frac{5}{8}$ 8

4 A kite is at a height of 16·37 metres. It descends in five equal drops of 0·04 metres. At what height is the kite now? ▢ m

5 Write the terms for each sequence.

a Count forwards in 0·6s from 3·5.

2nd term ▢   3rd term ▢   5th term ▢

b Count back in 0·04s from 16·57.

2nd term ▢   4th term ▢   5th term ▢

c Count back in $\frac{7}{8}$s from $18\frac{5}{8}$.

3rd term ▢   4th term ▢   6th term ▢

Date: _____   ☺ ☺ ☺

Number

Lesson 2: **Counting on and back beyond zero**

- Count on and back in steps of whole numbers and fractions from different numbers including negative numbers

1 Write the numbers in the sequence. Continue the count as far as you can.

a Count back in 2s.

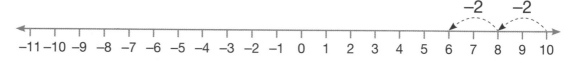

Sequence: 10, 8, 6, _____

b Count back in 0·2s.

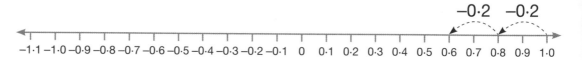

Sequence: 1, 0·8, 0·6, _____

c Count on in $\frac{3}{4}$s.

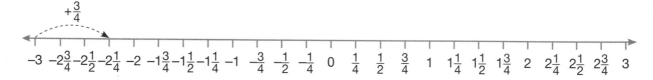

Sequence: $-3, -2\frac{1}{4}$, _____

  2 Complete the final bank balances in the table.

Start balance	$20	−$15	$1.40	−$2.30	−$3.10
Daily increase or decrease	Decrease $7 per day	Increase $4 per day	Decrease $0.30 per day	Increase $0.40 per day	Increase $0.80 per day
Number of days	4	9	6	7	8
Final bank balance					

3 Complete the missing numbers in the temperature table.

Start temperature (°C)	1·5	−2·3		−3·1	−2·7
Temperature rise/fall	**Fall** 0·4 degrees per hour	**Rise** 0·5 degrees per hour	**Fall** 0·7 degrees per hour	**Rise** 0·6 degrees per hour	**Rise** _____ degrees per hour
Number of hours	6	8	6		7
End temperature (°C)			−3·4	1·1	0·1

4 Maisie counts back from $1\frac{1}{10}$ in $\frac{3}{10}$s. Write the first six numbers Maisie will say in her counting sequence.

$1\frac{1}{10}$, ☐ , ☐ , ☐ , ☐ , ☐ , ☐

5 A camera is lowered into the sea in various stages. What depth does the camera reach after the following movements? (Heights above sea level are positive numbers; heights below are negative numbers.)

a The camera is lowered from a height of 1·3 metres. It is lowered in four stages of 0·6 metres and 3 stages of 0·7 metres.

Depth reached: ☐ m

b The camera is lowered from a height of 2·2 metres. It is lowered in five stages of 0·8 metres and four stages of 0·9 metres.

Depth reached: ☐ m

c The camera is lowered from a height of 2·5 metres. It is lowered in six stages of 0·4 metres, three stages of 0·7 metres and five stages of 0·9 metres.

Depth reached: ☐ m

Date: _____

Number

Lesson 3: **Finding the position-to-term rule**

• Know the rule of a sequence, and use the position of a term in the sequence to calculate its value

1 Use the rule for each sequence to generalise, complete the tables and answer the questions.

a Rule: multiply by 2

Position	1	2	3		5	6		8
Term	2		6	8			14	

i What are the next two terms in the sequence? ☐ and ☐

ii What is the position of the term with a value of 24? ☐

iii Robbie finds the term in position 24 by adding 24 + 20 + 4. Leona finds the term by calculating the sum of double 20 and double 4. Which method is more efficient? Why?

_____

b Rule: multiply by 4

Position	1	2	3	4	5	6	7	8
Term	4			16		24		

i What are the next two terms in the sequence? ☐ and ☐

ii What is the position of the term with a value of 44? ☐

2 Write the missing numbers in each table and answer the questions.

a

Position	1	2	3	4	5	6
Term	5	10	15	20		

i What is the position-to-term rule? _____

ii What is the value of the term in the 9th position? ☐

iii What is position of the term with a value of 100? ☐

Number

b

Position	1	2	3	4	5	6
Term	12	24	36	48		

i What is the position-to-term rule? _____

ii What is the value of the term in the 7th position? ☐

iii What is position of the term with a value of 120? ☐

 3 The positions and terms for a sequence have got mixed up.

a Draw lines to match the terms to their positions.

5   2   7   1   6   4   8   3

150   300   200   350   100   250   400   50

b What is the position-to-term rule? _____

c What is the value of the term in the 10th position? ☐

4 Solve the problems.

a Crates and the number of apples they contain are connected by a position-to-term rule. 4 crates have 100 apples, 5 crates have 125 apples and 6 crates have 150 apples. How many apples will there be in:

i 2 crates? ☐   **ii** 9 crates? ☐

iii 14 crates? ☐   **iv** 19 crates? ☐

b Jars and the number of beads they contain are connected by a position-to-term rule. 7 jars have 84 beads, 8 jars have 96 beads and 9 jars have 108 beads. How many beads will there be in:

i 4 jars? ☐   **ii** 12 jars? ☐

iii 16 jars? ☐   **iv** 23 jars? ☐

Date: _____

Lesson 4: **Finding terms of a square number sequence**

- Use the position of a term in the sequence of square numbers to calculate its value

1 Draw and write to complete the table.

Position	1	2	3	4	5
Pattern			(pattern)		(pattern)
Term		$2 \times 2 = 4$		$4 \times 4 = 16$	

2 Complete the table with all the square numbers up to 10^2.

Position	Calculation	Value
1	1^2	1
2	2^2	4
3		
4		
5		
6		
7		
8		
9		
10		

Number

3 Solve the following problems.

a What is the area of a square with sides 6 cm? ⬚ cm²

b The children in a playground arrange themselves in 9 rows of 9.

How many children are in the playground? ⬚

c 13 zebras each have 13 stripes. How many stripes is that

altogether? ⬚

4
a Can you specialise and find two square numbers that add together to make another square number? Write your answer and working out in the box below.

⬚

b Can you find three square numbers that add together to make another square number? Write your answer and working out in the box below.

⬚

c Alex solves the problems above by randomly adding square numbers together to see if they make another square number. Majeda starts by writing all the square numbers in a row. She thinks of each square number in turn (from 2^2 upwards) and looks for numbers in the row that sum to that square number.
Why do you think Majeda's method is more efficient than Alex's?

_____

_____

_____

Date: _____ 😊 😐 ☹

Number

Lesson 1: **Adding positive and negative numbers (1)**

- Use a number line to add positive and negative numbers

1 Estimate the answer and then use the number track to find the sum. How close was your estimate? Begin at the augend. Count on to the right the number of places given by the addend.

-10	-9	-8	-7	-6	-5	-4	-3	-2	-1	0	1	2	3	4	5	6	7	8	9	10

a $-4 + 3 =$ ☐

Estimate: ☐

b $-6 + 2 =$ ☐

Estimate: ☐

c $-8 + 6 =$ ☐

Estimate: ☐

d $-7 + 5 =$ ☐

Estimate: ☐

e $-9 + 8 =$ ☐

Estimate: ☐

f $-8 + 7 =$ ☐

Estimate: ☐

g $-8 + 14 =$ ☐

Estimate: ☐

h $-5 + 15 =$ ☐

Estimate: ☐

i $-9 + 18 =$ ☐

Estimate: ☐

j $-10 + 18 =$ ☐

Estimate: ☐

k $-9 + 19 =$ ☐

Estimate: ☐

l $-2 + 9 =$ ☐

Estimate: ☐

 2 Estimate the answer and then use the number line to find the sum. How close was your estimate?

-20	-19	-18	-17	-16	-15	-14	-13	-12	-11	-10	-9	-8	-7	-6	-5	-4	-3	-2	-1	0	1	2	3	4	5	6	7	8	9	10	11	12	13	14	15	16	17	18	19	20

a $-11 + 8 =$ ☐

Estimate: ☐

b $-14 + 12 =$ ☐

Estimate: ☐

c $-13 + 5 =$ ☐

Estimate: ☐

d $-15 + 14 =$ ☐

Estimate: ☐

e $-16 + 15 =$ ☐

Estimate: ☐

f $-14 + 13 =$ ☐

Estimate: ☐

g −17 + 28 = ☐   **h** −18 + 31 = ☐   **i** −19 + 29 = ☐

Estimate: ☐   Estimate: ☐   Estimate: ☐

j −20 + 39 = ☐   **k** −6 + 14 = ☐   **l** −9 + 19 = ☐

Estimate: ☐   Estimate: ☐   Estimate: ☐

 3 Calculate the new bank balance. Write the calculation.

Starting balance	Money in	New balance	Calculation
−$16	$23		−16 + 23 =
−$18	$26		
−$11	$24		
−$14	$32		
−$4	$35		
−$2	$37		

4 Write two numbers, a negative augend and a positive addend, that will give each total.

The negative augend must be greater than −10.

Example: ☐ −5 ☐ + ☐ 18 ☐ = 13

a ☐ + ☐ = −4   **b** ☐ + ☐ = −6

c ☐ + ☐ = −8   **d** ☐ + ☐ = −3

e ☐ + ☐ = −2   **f** ☐ + ☐ = −5

g ☐ + ☐ = 18   **h** ☐ + ☐ = 19

i ☐ + ☐ = −1   **j** ☐ + ☐ = 2

Date: _____

Number

Number

Lesson 2: **Adding positive and negative numbers (2)**

• Use a number scale to add positive and negative numbers

1 Estimate the answer and then use the thermometer to answer the question. How close was your estimate?

°C

a   What is 7 degrees more than −15°C?   ☐ °C Estimate: ☐

b   What is 9 degrees more than −20°C?   ☐ °C Estimate: ☐

c   What is 12 degrees more than −16°C?   ☐ °C Estimate: ☐

d   What is 10 degrees more than −8°C?   ☐ °C Estimate: ☐

e   What is 15 degrees more than −5°C?   ☐ °C Estimate: ☐

f   What is 25 degrees more than −15°C?   ☐ °C Estimate: ☐

2 Use the sea level scale to help you answer the questions. Remember, measurements below sea level are negative values. For example, 3 metres below sea level has the value −3 m.

a   A diver is 15 metres below sea level.

He rises 8 metres. What is his new depth? ☐ m

b   A fish rises 16 metres from a depth of −17 metres.

What is its new depth? ☐ m

c   A shark rises 18 metres from a depth of −19 metres.

What is its new depth? ☐ m

d   A crane lifts a treasure chest from a depth of −16 metres. If the treasure chest is lifted 29 metres, what is its

new position? ☐ m

e   A waterproof rocket rises 36 metres from the ocean floor from a depth of −18 metres.

What is its new position? ☐ m

Number

3 Estimate the answer and then complete the calculation. How close was your estimate?

a $-17 + 24 =$ [ ]

Estimate: [ ]

b $-19 + 26 =$ [ ]

Estimate: [ ]

c $-22 + 19 =$ [ ]

Estimate: [ ]

d $-23 + 26 =$ [ ]

Estimate: [ ]

e $-18 + 34 =$ [ ]

Estimate: [ ]

f $-24 + 42 =$ [ ]

Estimate: [ ]

g $-44 + 52 =$ [ ]

Estimate: [ ]

h $-52 + 67 =$ [ ]

Estimate: [ ]

i $-55 + 73 =$ [ ]

Estimate: [ ]

j $-58 + 84 =$ [ ]

Estimate: [ ]

k $-46 + 75 =$ [ ]

Estimate: [ ]

l $-37 + 83 =$ [ ]

Estimate: [ ]

4 Write the missing numbers in the table.

Temperature (°C)	Temperature rise (°C)	New temperature (°C)
−36	77	
−27	64	
−32		48
−24		62
	64	37
	75	33
	58	27

Date: _____

Number

Lesson 3: Identifying values for symbols in addition calculations

- Find the value of unknown values in calculations that are represented by symbols

1 Work out the unknown values.

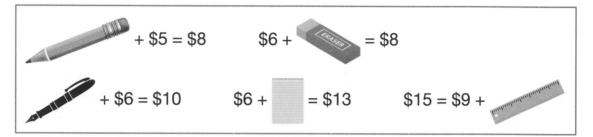

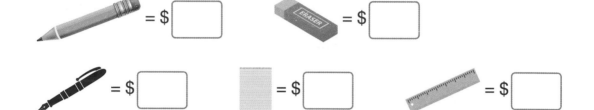

2 Work out the unknown values.

a $36 + a = 76$ $a =$ [ ]

b $b + 40 = 83$ $b =$ [ ]

c $c + 17 = 55$ $c =$ [ ]

d $43 + d = 69$ $d =$ [ ]

e $12 + e = 52$ $e =$ [ ]

f $f + 30 = 53$ $f =$ [ ]

g $73 = 29 + g$ $g =$ [ ]

h $h + 33 = 51$ $h =$ [ ]

i $41 + i = 89$ $i =$ [ ]

j $71 = j + 36$ $j =$ [ ]

k $k + 53 = 72$ $k =$ [ ]

l $57 + l = 95$ $l =$ [ ]

3 For each word problem, write a number sentence. Use a letter to represent the unknown value. Then solve the calculation to find the unknown value. Write your answer and working out in the box.

a The early morning temperature is 15°C. The temperature rises throughout the day. By the evening, it is 31°C. How much has the temperature risen?

b A shopping trolley and a large bag of shopping have a mass of 56 kg. The bag has a mass of 9 kg. What is the mass of the trolley?

c A large tank contains 63 litres of water. A tap above the tank is opened and more water is poured into the tank. The tank now has 92 litres of water. How much was poured into the tank?

4 Work out the unknown values.

a $133 + a = 224$ $a =$

b $b + 156 = 235$ $b =$

c $c + 143 = 351$ $c =$

d $234 + d = 463$ $d =$

e $467 + e = 524$ $e =$

f $f + 376 = 611$ $f =$

g $335 = 177 + g$ $g =$

h $h + 566 = 723$ $h =$

i $654 + i = 812$ $i =$

j $813 = j + 687$ $j =$

k $k + 765 = 921$ $k =$

l $856 + l = 923$ $l =$

Date: _____

Number

Lesson 4: **Identifying values for symbols in subtraction calculations**

- Find the value of unknown values in calculations that are represented by symbols

1 Work out the unknown values.

a

[ ] ml   400 ml

b

600 ml   350 ml

c

420 ml   765 ml

d

785 ml   920 ml

2 Work out the unknown values.

a $65 - a = 32$ $a =$ [ ]

b $b - 27 = 24$ $b =$ [ ]

c $c - 18 = 43$ $c =$ [ ]

d $57 - d = 18$ $d =$ [ ]

e $71 - e = 36$ $e =$ [ ]

f $f - 59 = 28$ $f =$ [ ]

g $48 = 72 - g$ $g =$ [ ]

h $h - 44 = 32$ $h =$ [ ]

i $94 - i = 57$ $i =$ [ ]

j $27 = j - 66$ $j =$ [ ]

k $k - 65 = 76$ $k =$ [ ]

l $121 - l = 69$ $l =$ [ ]

3 Work out the secret numbers.

a Leo says: '58 less than my secret number is 33.'
What is Leo's secret number? [ ]

b Filipa says: '96 minus my secret number is 18.'
What is Filipa's secret number? [ ]

Number

4 For each word problem, write a number sentence. Use a letter to represent the unknown value. Then solve the calculation to find the unknown value. Write your answer and working out in the box.

a The temperature falls by 17 degrees throughout the day. By the evening, the temperature is 6 °C. What was the temperature in the morning?

b Lana has $72 to spend. She buys a video game and is left with $17. What was the price of the video game?

c Charlie is left with 86 stickers after giving some away. He had 142 stickers to begin with. How many did he give away?

5 Work out the unknown values.

a $246 - a = 178$ $a =$

b $b - 197 = 76$ $b =$

c $c - 188 = 165$ $c =$

d $312 - d = 186$ $d =$

e $421 - e = 255$ $e =$

f $f - 243 = 229$ $f =$

g $473 = 625 - g$ $g =$

h $h - 448 = 377$ $h =$

i $867 - i = 594$ $i =$

j $294 = j - 618$ $j =$

k $k - 264 = 677$ $k =$

l $915 - l = 693$ $l =$

Date: _____

Number

Lesson 1: **Subtracting positive and negative integers**

• Find the difference between positive and negative numbers

1 Use the number track to find the difference between the two numbers. Begin at the lower number and count on to the higher number.

−10	−9	−8	−7	−6	−5	−4	−3	−2	−1	0	1	2	3	4	5	6	7	8	9	10

a −2 and 2 ☐ **b** 4 and −1 ☐ **c** −3 and 2 ☐

d −7 and 5 ☐ **e** 5 and −2 ☐ **f** −5 and 3 ☐

g −9 and 7 ☐ **h** −10 and 6 ☐ **i** 10 and −6 ☐

j −8 and 9 ☐ **k** −10 and 10 ☐ **l** 6 and −2 ☐

2 Cai reads the temperature on a thermometer in a glass of frozen water. The temperature is −6 °C. He leaves the water by a radiator and takes the temperature a few hours later. It is 12 °C. What is the difference between the two temperature readings? ☐ degrees

3 Estimate the answer and then use the number line to find the difference between the two numbers. How close was your estimate?

−20 −19 −18 −17 −16 −15 −14 −13 −12 −11 −10 −9 −8 −7 −6 −5 −4 −3 −2 −1 0 1 2 3 4 5 6 7 8 9 10 11 12 13 14 15 16 17 18 19 20

a −11 and 2 ☐ **b** −3 and 12 ☐ **c** 13 and −4 ☐
Estimate: ☐ Estimate: ☐ Estimate: ☐

d −5 and 11 ☐ **e** 14 and −5 ☐ **f** −6 and 15 ☐
Estimate: ☐ Estimate: ☐ Estimate: ☐

g −16 and 14 ☐ **h** 17 and −13 ☐ **i** −17 and 16 ☐
Estimate: ☐ Estimate: ☐ Estimate: ☐

j 18 and −19 ☐   **k** 11 and −13 ☐   **l** −19 and 12 ☐

Estimate: ☐   Estimate: ☐   Estimate: ☐

 Calculate the difference between the starting temperature and the final temperature.

Starting temperature	Final temperature	Temperature change
−13 °C	9 °C	degrees
14 °C	−12 °C	degrees
−15 °C	11 °C	degrees
17 °C	−8 °C	degrees
−16 °C	13 °C	degrees
19 °C	−17 °C	degrees

5 Write a number to make the difference given. Write a positive number if the first number is negative. Write a negative number if the first number is positive.

a 13: −2 and ☐   **b** 15: 13 and ☐   **c** 14: −6 and ☐

d 17: 11 and ☐   **e** 12: −8 and ☐   **f** 16: 13 and ☐

g 19: −8 and ☐   **h** 18: 4 and ☐   **i** 23: −12 and ☐

j 21: 9 and ☐   **k** 24: −3 and ☐   **l** 24: 2 and ☐

6 Compare the method you used to solve the questions in **5** with a classmate. Which method is the most efficient? Why?

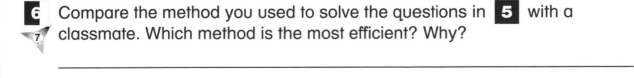

Date: _____

Number

Lesson 2: **Subtracting two negative integers**

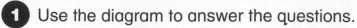

- Find the difference between two negative numbers

1 Use the diagram to answer the questions.

a What is 1 metre deeper than –6 m? ☐ m

4 **b** Why is the answer to part **a** not –5?

_____

c What is 2 metres deeper than –3 m? ☐ m

d How many metres are there between these depths:

i –2 m and –4 m? ☐ m

iii –4 m and –9 m? ☐ m

ii –3 m and –6 m? ☐ m

iv –5 m and –7 m? ☐ m

2 Estimate each answer and then use the number line to find the difference between the two numbers. How close was your estimate?

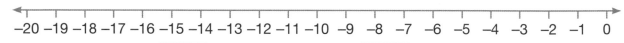

−20 −19 −18 −17 −16 −15 −14 −13 −12 −11 −10 −9 −8 −7 −6 −5 −4 −3 −2 −1 0

a −5 and −2 ☐

Estimate: ☐

b −3 and −7 ☐

Estimate: ☐

c −8 and −4 ☐

Estimate: ☐

d −5 and −9 ☐

Estimate: ☐

e −8 and −1 ☐

Estimate: ☐

f −6 and −10 ☐

Estimate: ☐

g −13 and −8 ☐

Estimate: ☐

h −13 and −18 ☐

Estimate: ☐

i −12 and −3 ☐

Estimate: ☐

j −9 and −18 ☐

Estimate: ☐

k −17 and −14 ☐

Estimate: ☐

l −7 and −20 ☐

Estimate: ☐

3 Answer the word problems.

a The temperature in the morning is −4 °C. By the evening, it has fallen to −9 °C. By how many degrees has the temperature fallen?

☐ degrees

Number

b Ryan has –$4 in his bank account. He buys a book and checks his account. It is now –$16. What was the price of the book? ☐

c A fish swims from a depth of –19 metres to a depth of –3 metres. How far did it swim? ☐ m

d A lift descends from basement level –4 to level –20. How many floors is that? ☐

 4 Estimate each answer and then work out the difference between the two numbers. How close was your estimate?

a –17 and –24 ☐

Estimate: ☐

b –26 and –19 ☐

Estimate: ☐

c –17 and –25 = ☐

Estimate: ☐

d –28 and –16 ☐

Estimate: ☐

e –29 and –14 ☐

Estimate: ☐

f –31 and –18 = ☐

Estimate: ☐

g –16 and –43 ☐

Estimate: ☐

h –51 and –27 ☐

Estimate: ☐

i –33 and –72 = ☐

Estimate: ☐

5 Each day, the crew of a submarine practise raising the submarine. A record is kept of the depths involved. Complete the missing numbers in the table.

	Lower depth (m)	Upper depth (m)	Height moved (m)
Monday	–52	–16	
Tuesday	–63	–18	
Wednesday	–55		48
Thursday	–66		37
Friday		–13	58
Saturday		–14	77

Date: _____

Lesson 3: **Identifying values of variables in calculations (1)**

• Identify the values of variables in calculations

1 6 cubes are fixed together. The cubes can be green or yellow.

What numbers of green and yellow cubes are possible?

Complete the missing numbers to show all the possible combinations.

☐ yellow and ☐ green   ☐ yellow and ☐ green

☐ yellow and ☐ green   ☐ yellow and ☐ green

☐ yellow and ☐ green

2 If $p + q = 7$, and p and q are whole numbers, what are all the possible solutions for the values of p and q? Complete the table of solutions.

p								
q								

3 If $m + n = 20$, and m and n are both whole numbers from 7 and 13, what are all the possible solutions for the values of m and n? Complete the table of solutions.

m						
n						

4 Write an equation that describes the relationship between the two values described. In each question, find all the possible solutions. Assume all numbers are whole numbers.

a The sum of two numbers is 14. Both numbers are from 3 and 11.

Equation:

Solutions:

b The sum of two numbers is 22. Both numbers are from 6 and 16.

Equation:

Solutions:

c Two metal weights have a total mass of 18 kg. Both weights are from 4 kg and 14 kg.

Equation:

Solutions:

5 Write an equation that describes the relationship between the two values described. In each question, find all the possible solutions. Assume all numbers are whole numbers.

a The difference between two numbers is 7. Both numbers are from 25 and 37.

Equation:

Solutions:

b Variables x and y are connected by the statement '2 times x plus 5 equals y'. Both x and y are from 3 and 19.

Equation:

Solutions:

Date: _____

Number

Lesson 4: **Identifying values of variables in calculations (2)**

• Use a simple formula for given values

1 The cost to rent a rowing boat at the park is shown by the formula:

Cost in dollars = 14 + n and n is the number of people.

How much does it cost for these numbers of people to rent a boat?

a 3

Cost in dollars is 14 + ☐ = ☐

b 4

Cost in dollars is ☐ + ☐ = ☐

c 6

Cost in dollars is ☐

2 Susie works in a restaurant. She is paid $95 per day, plus whatever she receives in tips.

The amount she earns each day is given by the formula E = $95 + T where E is amount earned and T is tips.

Use the formula to complete the table.

T ($)	9	15	23	38	46	57
E ($)						

3 The perimeter of a square is given by the formula P = s + s + s + s

where P is perimeter and s is the length of the sides.

Use the formula to complete the table.

Side length (cm)	7	14	23	37	58	77
Perimeter (cm)						

Number

4 The area of a square is given by the formula $A = s^2$

where A is area and s is the length of the sides.

Use the formula to complete the table.

Side length (cm)	4	6	7	9	12	20
Area (cm²)						

5 $b - a = 17$

Use the equation to complete the table.

a		8		16	23	39
b	23		35			

6 Adil and Adnan are brothers. Adil is older than Adnan and gets more pocket money.

The amount of pocket money Adnan receives is given by the formula $H = W - \$17$ where H is the amount of money Adnan gets and W is the amount Adil gets.

Use the formula to complete the table.

W ($)	42		63		76	
H ($)		57		84		95

7 Write the formula for the perimeter of a rectangle where l is length and w is width.

Use the formula to calculate all the whole number solutions for l and w for a rectangle that has a perimeter of:

a 8 cm

b 12 cm

c 20 cm

Date: _____

Lesson 1: **Common multiples**

Number

• Understand and find common multiples

1 Write the first ten multiples of each number.

a Multiples of 4

× 1	× 2	× 3	× 4	× 5	× 6	× 7	× 8	× 9	× 10

b Multiples of 6

× 1	× 2	× 3	× 4	× 5	× 6	× 7	× 8	× 9	× 10

c Multiples of 9

× 1	× 2	× 3	× 4	× 5	× 6	× 7	× 8	× 9	× 10

2 List the first 15 multiples of each number, then write the common multiples of both numbers.

a Multiples of 4: ☐ ☐ ☐ ☐ ☐ ☐ ☐
☐ ☐ ☐ ☐ ☐ ☐ ☐ ☐

Multiples of 5: ☐ ☐ ☐ ☐ ☐ ☐ ☐
☐ ☐ ☐ ☐ ☐ ☐ ☐ ☐

The common multiples of 4 and 5 include: ☐

b Multiples of 2: ☐ ☐ ☐ ☐ ☐ ☐ ☐
☐ ☐ ☐ ☐ ☐ ☐ ☐ ☐

Number

Multiples of 5: ☐ ☐ ☐ ☐ ☐ ☐ ☐

☐ ☐ ☐ ☐ ☐ ☐ ☐

The common multiples of 2 and 5 include: ☐

3 Leah and Logan place tins on a shelf.

Leah places tins in stacks of 3. Logan places them in stacks of 10.

They stack the same number of tins.

a What is the smallest number of tins each could have stacked? ☐

Show your working out in the box.

b Compare the method you used to solve this problem with a classmate.
Which method is the most efficient? Why?

_____

_____

_____

4 Find the first two common multiples of each pair of numbers.

a 7 and 12: ☐ and ☐　　**b** 9 and 11: ☐ and ☐

c 8 and 13: ☐ and ☐　　**d** 12 and 13: ☐ and ☐

Date: _____

Lesson 2: **Common factors**

> • Understand and find common factors

1 Find the factors of these numbers.

a 16: ☐ ☐ ☐ ☐ ☐

b 24: ☐ ☐ ☐ ☐ ☐ ☐ ☐ ☐

c 30: ☐ ☐ ☐ ☐ ☐ ☐ ☐ ☐

d 36: ☐ ☐ ☐ ☐ ☐ ☐ ☐ ☐ ☐

e 50: ☐ ☐ ☐ ☐ ☐ ☐

f 34: ☐ ☐ ☐ ☐

2 Write the factors in the correct part of each Venn diagram. The common factors go in the intersection (where the ovals overlap).

Example:

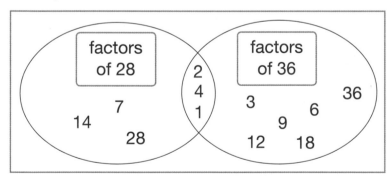

a

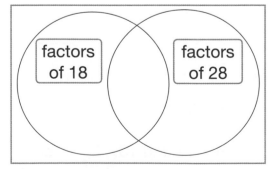

b

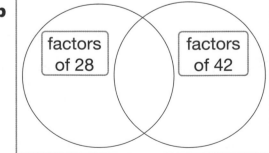

Number

c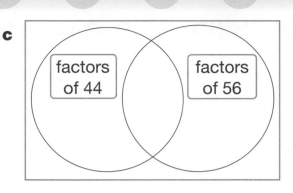

factors of 44   factors of 56

d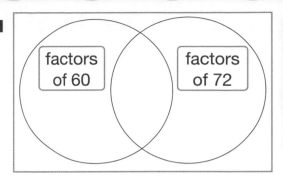

factors of 60   factors of 72

3 Murray has a bag of 36 red beads and Grace has a bag of 44 yellow beads.

They divide the beads into trays so that each tray has the same number of red and yellow beads.

What is the largest number of trays that can be used?

Show your working out in the box.

4 Find all the common factors of each pair of numbers.

a 66 and 84:

b 76 and 92:

c 66 and 99:

d 84 and 108:

Date: _____

Lesson 3: **Tests of divisibility by 3, 6 and 9**

- Test numbers for divisibility by 3, 6 and 9

1 Write the missing numbers.

a $2 \times 6 = \boxed{}$ **b** $3 \times 9 = \boxed{}$ **c** $8 \times 3 = \boxed{}$

d $9 \times 5 = \boxed{}$ **e** $6 \times 3 = \boxed{}$ **f** $3 \times 7 = \boxed{}$

g $3 \times \boxed{} = 18$ **h** $\boxed{} \times 5 = 45$ **i** $7 \times \boxed{} = 63$

j $\boxed{} \times 6 = 36$ **k** $6 \times \boxed{} = 42$ **l** $\boxed{} \times 4 = 36$

2 Draw a ring around the numbers that are divisible by 2.

633   447   144   587   454   986

346   105   888   711   432   549

3 **a** Draw a ring around the numbers that are divisible by 3.

469   509   293   981   890   228   423   548

633   831   127   160   483   834   201   758

b Draw a ring around the numbers that are divisible by 6.

220   367   771   953   184   725   445   595

322   673   131   190   771   929   667   636

c Draw a ring around the numbers that are divisible by 9.

578   459   794   310   181   417   879   570

399   822   314   550   280   595   931   746

Number

4 Write a 3-digit number that is:

  a divisible by 6 with a tens digit of 5

  b divisible by 9 with a ones digit of 4

  c divisible by 6 with a hundreds digit of 5

  d divisible by 3, 6 and 9 with a tens digit of 2

5 Write a 4-digit number that is:

  a divisible by 6 with a hundreds digit of 7

  b divisible by 3 with a tens digit of 8

  c divisible by 9 with a ones digit of 8

  d divisible by 3 only with a hundreds digit of 9

6 Work out the secret numbers.

  a Maha says: 'My secret number is a 4-digit number that is divisible by 3, 6 and 9. It has a hundreds digit of 7 and a ones digit of 4.'

  What is Maha's secret number?

  b April says: 'My secret number is a 4-digit number that is divisible by 3, 6 and 9. It has a thousands digit of 7 and a tens digit of 1.'

  What is April's secret number?

  c Tim says: 'My secret number is a 4-digit number that is divisible by 3, 6 and 9. It has a thousands digit of 6 and a ones digit of 6.'

  What is Tim's secret number?

Date: _____

Number

Lesson 4: **Cube numbers**

• Use square numbers to recognise cube numbers

1 Complete the multiplications. Show your working out in the box.

a $2 \times 2 \times 2 = $ ☐   **b** $4 \times 4 \times 4 = $ ☐   **c** $1 \times 1 \times 1 = $ ☐

d $3 \times 3 \times 3 = $ ☐   **e** $5 \times 5 \times 5 = $ ☐

2 What is the value of each of these?

a 3^3 ☐   **b** 1^3 ☐   **c** 5^3 ☐

d 2^3 ☐   **e** 4^3 ☐

3 Complete the statements.

a The cube of 1 or 1^3 is 1 . 1 is a cube number.

b The cube of 4 or ☐ is ☐ . ☐ is a cube number.

c The cube of 2 is ☐ . ☐ is a cube number.

d The cube of 5 is ☐ . ☐ is a cube number.

e The cube of 3 is ☐ . ☐ is a cube number.

4 Which of the following do **not** represent a cube number? Cross them out.

4^3 $3^2 \times 9$ $5 \times 5 \times 25$ $2^2 \times 2$

16×4^2 64 $1 \times 1 \times 1$ 27

5 Complete the missing numbers.

a $3 \times 3 \times $ ☐ $= 27$ **b** ☐ $\times 5^2 = 125$ **c** $4^2 \times 4 = $ ☐

d $\boxed{}^2 \times 2 = 8$ **e** $6^2 \times 6 = \boxed{}$ **f** $8 \times 8^2 = \boxed{}$

g $\boxed{} \times 10^2 = 1000$ **h** $9^2 \times 9 = \boxed{}$ **i** $7 \times 7^2 = \boxed{}$

6 Solve the problems. Use the space at the bottom for working out.

a Tooka makes 5 necklaces from beads. Each necklace has 5 beads.

How many beads does she use in total? $\boxed{}$

b Alfie took a number and cubed it to get 64.

What was the number he cubed? $\boxed{}$

c The square of a number is 9. What is the cube of the number? $\boxed{}$

d The cube of a number is 8. What is the square of the number? $\boxed{}$

e The square of a number is 100.

What is the cube of the number? $\boxed{}$

f The cube of a number is 729.

What is the square of the number? $\boxed{}$

g A model aircraft is made from 7 pieces.

How many pieces will be used to make 49 model aircraft? $\boxed{}$

h The mass of a brick is 8 kg. What is the mass of 64 bricks? $\boxed{}$ kg

Date: _____

Number

Number

Lesson 1: **Simplifying calculations (1)**

- Simplify calculations using the properties of number

1 Write the missing numbers in the calculations.

a $4 \times 13 \times 5 = \boxed{} \times 4 \times 13$ b $18 + 17 + 32 = 18 + \boxed{} + 17$

c $8 \times 67 = (8 \times \boxed{}) + (8 \times 7)$ d $50 \times 13 \times 2 = \boxed{} \times 50 \times 13$

2 Simplify and solve each calculation. Name the property of number you use to simplify.

a $23 + 14 + 27 + 36 = \boxed{} = \boxed{}$

Property: $\boxed{}$

b $8 \times 6 \times 5 \times 3 = \boxed{} = \boxed{}$

Property: $\boxed{}$

c $8 \times 38 + 7 = \boxed{} = \boxed{}$

Property: $\boxed{}$

d $15 \times 7 \times 6 + 5 = \boxed{} = \boxed{}$

Property: $\boxed{}$

e $47 + 28 + 133 + 152 = \boxed{} = \boxed{}$

Property: $\boxed{}$

f $13 + 7 \times 66 = \boxed{} = \boxed{}$

Property: $\boxed{}$

3 For each property of number, write a calculation that can be simplified using that property and then solve it.

Commutative

Solution:

Associative

Solution:

Distributive

Solution:

4 Simplify and solve.

a $25 \times 13 \times 8 \times 7 + 48 =$ ___ = ___

b $8 \times 57 + 7 \times 63 + 56 =$ ___ = ___

c $236 + 437 + 662 + 364 + 213 + 118 =$ ___ = ___

d $2128 + 8 \times 234 =$ ___ = ___

Date: _____

Number

Lesson 2: **Simplifying calculations (2)**

• Simplify calculations using the properties of number

1 Write a calculation for each problem. Simplify and solve it.

a A painter decorates 13 identical walls of a house. To paint each wall, she uses 4 cans of paint. How many cans of paint is that in total?

	=		=	

b A bag contains 13 red beads, 8 blue beads, 7 green beads and 12 yellow beads. How many beads is that in total?

	=		=	

2 Write a calculation for each problem. Simplify and solve it.

a There are 8 rows of 17 benches in the park. Each bench seats 5 people. How many people in total can be seated in the park?

	=		=	

b Mr Baxter counts the number of cars in a car park. 67 are red, 54 are green, 43 are black and 76 are blue. How many cars does Mr Baxter count?

	=		=	

c Jade provides seed to feed the hens in the hen houses on her farm. There are 20 hen houses. Jade uses 67 packets of seed per house. How many packets of seed does she use in total?

	=		=	

d 4 flocks of birds settle on each tree in an orchard. There are 8 trees and 150 birds in each flock. How many birds are in the orchard in total?

	=		=	

Number

3 Write a calculation for each problem. Simplify and solve it.

a Pots contain 77 pencils each. There are 6 pots on each table and 5 tables in total. An extra table has 38 pencils. How many pencils is that altogether?

[ ] = [ ] = [ ]

b A train has 6 carriages. The number of passengers travelling in the carriages are: 145, 258, 323, 112, 225, 117. How many passengers are there in total?

[ ] = [ ] = [ ]

4 Write a calculation for each problem. Simplify and solve it.

a There are 8 entrances on one side of a sports stadium. 236 people pass through each entrance.

There are 7 entrances on the other side of the stadium. 377 people pass through each one.

How many people are in the stadium in total?

[ ] = [ ] = [ ]

b Trays contain 133 toy bricks. There are 5 trays in each shelving unit and 8 shelving units in total. An extra tray has 260 bricks.

How many bricks is that altogether?

[ ] = [ ] = [ ]

c A lorry transports crates of potatoes from farms to supermarkets.

The mass of the crates in the lorry are: 1235 kg, 625 kg, 1414 kg, 336 kg, 1140 kg, 760 kg.

What is the combined mass of potatoes carried in the lorry?

[ ] = [ ] = [ ]

Date: _____

Lesson 3: **Using brackets (1)**

- Understand that when completing a calculation that includes brackets, operations in brackets must be completed first

Number

1 Complete the missing numbers.

You will need
- paper for working out

a $(4 + 2) \times 2 = \boxed{} \times 2 = \boxed{}$

b $10 \div (7 - 2) = 10 \div \boxed{} = \boxed{}$

c $(14 - 5) \times 5 = \boxed{} \times 5 = \boxed{}$

d $5 \times (15 + 7) = 5 \times \boxed{} = \boxed{}$

e $(57 + 13) \div 7 = \boxed{} \div 7 = \boxed{}$

f $(92 - 26) \times 3 = \boxed{} \times 3 = \boxed{}$

2 Simplify and solve each calculation.

a $(3 + 4) \times 5 = \boxed{} = \boxed{}$

b $6 \div (4 - 2) = \boxed{} = \boxed{}$

c $(9 - 6) \times 9 \boxed{} = \boxed{}$

d $7 \times (8 + 3) \boxed{} = \boxed{}$

e $(6 + 24) \div 3 = \boxed{} = \boxed{}$

f $15 \times (7 - 3) = \boxed{} = \boxed{}$

3 Simplify and solve each calculation.

a $(13 + 5) \times 7 = \boxed{} = \boxed{}$

b $96 \div (31 - 28) = \boxed{} = \boxed{}$

c $(42 - 35) \times 14 = \boxed{} = \boxed{}$

Number

d $24 \times (7 + 6) =$ [_____] = [__]

e $(58 + 38) \div 6 =$ [_____] = [__]

f $37 \times (71 - 26) =$ [_____] = [__]

4 Write your own calculations to simplify and solve.

a ([__] + [__]) × [__] = [_____] = [__]

b [__] ÷ ([__] − [__]) = [_____] = [__]

c ([__] − [__]) × [__] = [_____] = [__]

d [__] × ([__] + [__]) = [_____] = [__]

e ([__] + [__]) ÷ [__] = [_____] = [__]

5 Simplify and solve.

a $(18 + 37) \times 66 =$ [_____] = [__]

b $(354 - 163) \times 36 =$ [_____] = [__]

c $(346 + 146) \div 6 =$ [_____] = [__]

d $776 \div (803 - 795) =$ [_____] = [__]

6 We know that letters can be used as substitutes for numbers. If $a = 10$, $b = 7$ and $c = 5$, investigate what answers you get from the following. Use paper for working out if you need to.

a $(a + b) - c$ **b** $a + b + c$ **c** $a \div (b - c)$

d $a - (b + c)$ **e** $(a - b) + c$ **f** $(a + b) \times c$

g $a \times (b - c)$ **h** $(a - b) \times c$ **i** $a \times b - c$

j $a \times (b + c)$

Date: _____

☺ ☻ ☹

Lesson 4: **Using brackets (2)**

- Understand that when completing a calculation that includes brackets, operations in brackets must be completed first

1 **a** Mai makes 16 necklaces, each with 12 beads.

Unfortunately, she finds that 5 beads on each necklace are slightly cracked and she removes them.

How many beads, in total, are left on the necklaces?

Draw a ring around the calculation that correctly represents the problem.

$16 \times 12 - 5$ $(16 \times 12) - 5$ $12 - 5 \times 16$ $16 \times (12 - 5)$

 b Ling says that the answer is $16 \times 12 - 5 = 192 - 5 = 187$. What mistake has she made?

_____

2 Write a calculation for each problem. Simplify and solve it. Use brackets to show which part of the calculation needs to be solved first.

a 7 flowers grow in one part of a garden. 8 butterflies land on each flower. After 5 minutes, 3 butterflies fly away from each flower. How many butterflies are left?

[   ] = [   ] = [   ]

b Parcels contain 6 magazines and 4 books. How many parcels can be made from 290 magazines and books?

[   ] = [   ] = [   ]

c 15 athletes each run 12 km. Each athlete then runs a further 9 km. What is the combined distance run of all the athletes?

[   ] = [   ] = [   ]

d There are 7 trays on a table. 4 trays are removed. 2331 counters are then divided equally between the remaining trays. How many counters are in each tray?

[   ] = [   ] = [   ]

3 Liam went on a treasure hunt in his garden. He found 243 gold stars. Later he found 162 more. He then put the stars in piles of 5.

How many piles are there? ☐

Show your working out in the box.

4 Sarah planted 350 seeds. She shared them equally between 7 different garden plots. In one plot, 13 seeds did NOT grow into flowers.

How many seeds in this plot did grow into flowers? ☐

Show your working out in the box.

5 Each large sack in a warehouse contains 123 kg of potatoes. Sacks are arranged in groups of 5 and there are 17 groups. 48 kg of potatoes are removed from each sack.

What amount of potatoes is left in the sacks? ☐

Show your working out in the box.

Date: _____

Lesson 1: **Multiplying by 1-digit numbers (1)**

Number

• Use the expanded written method to multiply numbers up to 10 000 by 1-digit whole numbers

You will need
• paper for working out

1 Round and estimate first, then multiply.

a Estimate: [   ]

			7	6
×				4
+				

	×	
	×	

b Estimate: [   ]

			5	8
×				8
+				

	×	
	×	

c Estimate: [   ]

		2	3	4
×				3
+				

	×	
	×	

d Estimate: [   ]

		4	5	6
×				5
+				

	×	
	×	

 2 Round and estimate first, then multiply.

a Estimate: [   ]

	3	6	2	5
×				3
+				

	×	
	×	
	×	
	×	

b Estimate: [   ]

	6	4	3	3
×				4
+				

	×	
	×	
	×	
	×	

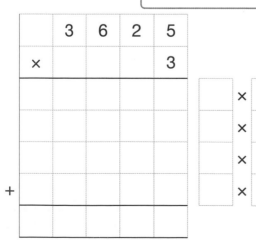

c Estimate: [   ]

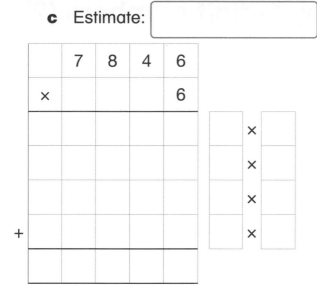

	7	8	4	6
×				6
+				

× [ ]
× [ ]
× [ ]
× [ ]

d Estimate: [   ]

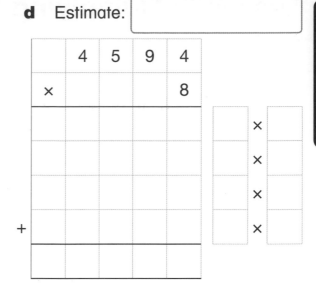

	4	5	9	4
×				8
+				

× [ ]
× [ ]
× [ ]
× [ ]

 3 Solve each word problem using the expanded written method. Show your working in the boxes.

a A company receives 4 payments of $2467. How much money has the company received in total?

Estimate: [   ]

[   ]

Total received: $ [   ]

b A large stadium is divided into 6 areas. Each area has 7347 seats. How many seats is that altogether?

Estimate: [   ]

[   ]

Total number of seats:

[   ]

4 Order the trains by the distance travelled, from shortest to longest.

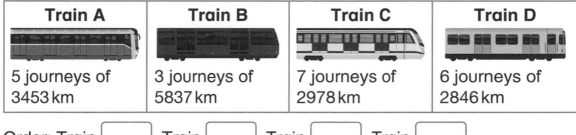

Train A	Train B	Train C	Train D
5 journeys of 3453 km	3 journeys of 5837 km	7 journeys of 2978 km	6 journeys of 2846 km

Order: Train [   ], Train [   ], Train [   ], Train [   ]

Date: _____

Number

Lesson 2: **Multiplying by 1-digit numbers (2)**

- Use the formal written method to multiply numbers up to 10 000 by 1-digit whole numbers

1 Round and estimate first, then multiply.

a Estimate:

	1	2	4	3
×				3
+				

×
×
×
×

b Estimate:

	5	4	2	4
×				4
+				

×
×
×
×

c Estimate:

	4	6	5	6
×				6
+				

×
×
×
×

d Estimate:

	7	8	6	7
×				8
+				

×
×
×
×

Number

 2 Use the formal written method to calculate. Round and estimate first, then multiply.

a Estimate: [ ]

	2	4
×		6

b Estimate: [ ]

	7	3
×		8

c Estimate: [ ]

	5	6	7
×			6

3 Solve each multiplication using the formal written method. Set out your working in the boxes.

a 3464 × 3 =

Estimate: [ ]

b 7475 × 4 =

Estimate: [ ]

c 6358 × 7 =

Estimate: [ ]

4 Captain Smith takes his air balloon to a height of 2326 metres. He does this three times in Week 1 (6 journeys in total, up and down). In Week 2, he takes the air balloon to a height of 1742 metres. He does this four times in the week (8 journeys in total, up and down). In which week does the air balloon travel the furthest distance?

Show your working in the box. Use the formal written method.

Week [ ]

Date: _____

Lesson 3: **Multiplying by 2-digit numbers (1)**

- Use partitioning and the grid method to multiply numbers up to 10 000 by 2-digit numbers

1 For each calculation, use the grid method to work out the answer.

a 48 × 27 = ⬜

× ⬜ ⬜

⬜ ⬜ ⬜
⬜ ⬜ ⬜

b 86 × 54 = ⬜

× ⬜ ⬜

⬜ ⬜ ⬜
⬜ ⬜ ⬜

2 Simplify, then use partitioning to multiply the numbers.

a 3734 × 40 = 3734 × 4 × 10

3734 × 4 = (⬜ × 4) + (⬜ × 4) + (⬜ × 4) + (⬜ × 4)

= ⬜ + ⬜ + ⬜ + ⬜

= ⬜

3734 × 40 = ⬜ × 10

= ⬜

b 5859 × 60 = 5859 × ⬜ × 10

5859 × 6 = (⬜ × 6) + (⬜ × 6) + (⬜ × 6) + (⬜ × 6)

= ⬜ + ⬜ + ⬜ + ⬜

= ⬜

5859 × 60 = ⬜ × 10

= ⬜

Number

3 $8342 is put into a bank account.

a The amount increases in value and after 10 years, it is worth 30 times more. How much is in the account? Show your working in the box.

b 20 years after the amount was first put into the account, it is worth 67 times more. How much is in the account? Show your working in the box.

4 For each calculation, estimate first, then use the grid method to work out the answer.

a 452 × 67 = [ ]

Estimate: [ ]

b 6829 × 58 = [ ]

Estimate: [ ]

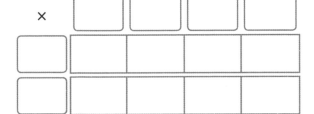

Date: _____

Lesson 4: **Multiplying by 2-digit numbers (2)**

Number

- Use the formal written method of long multiplication to multiply numbers up to 10 000 by 2-digit numbers.

You will need
- squared paper

1 Round and estimate first, then multiply using the expanded written method.

a Estimate: [ ]

			5	8
×			4	6
+				

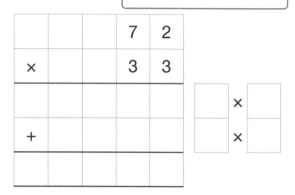

b Estimate: [ ]

			7	2
×			3	3
+				

c Estimate: [ ]

		2	1	3
×			2	4
+				

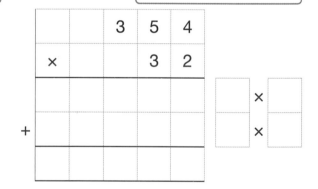

d Estimate: [ ]

		3	5	4
×			3	2
+				

2 Calculate using the formal written method (long multiplication).

a

		3	3
×		2	4
			0

b

		5	3
×		3	5
			0

c

		6	3
×		4	6
			0

Number

d

	1	6	3
×		3	2
			0

e

	3	4	3
×		2	8
			0

f

	4	5	4
×		3	6
			0

3 Calculate using the formal written method (long multiplication).

a

		2	3	4	6
×				1	7

b

		1	4	2	3
×				2	8

c

		2	3	4	7
×				3	5

d

		4	7	2	8
×				5	6

4 Round and estimate first, then use squared paper to solve each calculation using long multiplication.

a 474 × 74 = [　　]

Estimate: [　　]

b 786 × 88 = [　　]

Estimate: [　　]

c 3467 × 47 = [　　]

Estimate: [　　]

d 8179 × 23 = [　　]

Estimate: [　　]

Date: _____

Number

Lesson 1: **Dividing 2-digit numbers by 1-digit numbers (1)**

- Use the 'chunking' method to divide 2-digit numbers by 1-digit numbers

1 Write the missing numbers and find the quotient. Example:

$72 \div 4 = (\boxed{40} \div 4) + (\boxed{32} \div \boxed{4}) = \boxed{10} + \boxed{8} = \boxed{18}$

a $81 \div 3 = (\boxed{} \div 3) + (\boxed{} \div 3) = \boxed{} + \boxed{} = \boxed{}$

b $96 \div 6 = (\boxed{} \div 6) + (\boxed{} \div 6) = \boxed{} + \boxed{} = \boxed{}$

c $91 \div 7 = (\boxed{} \div 7) + (\boxed{} \div 7) = \boxed{} + \boxed{} = \boxed{}$

d $98 \div 7 = (\boxed{} \div 7) + (\boxed{} \div 7) = \boxed{} + \boxed{} = \boxed{}$

2 Use place value counters to model each division calculation, crossing out counters when you regroup. Estimate the answer first. Complete the steps to find the quotient and write the remainder as a whole number.

a $99 \div 4$　　Estimate: $\boxed{}$　　　　$99 \div 4 = \boxed{}$ r $\boxed{}$

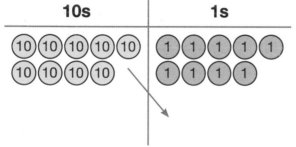

b $89 \div 6$　　Estimate: $\boxed{}$　　　　$89 \div 6 = \boxed{}$ r $\boxed{}$

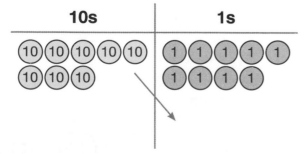

c 93 ÷ 7   Estimate: ⬚   93 ÷ 7 = ⬚ r ⬚

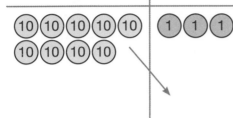

10s	1s

10s	1s

d 97 ÷ 8   Estimate: ⬚   97 ÷ 8 = ⬚ r ⬚

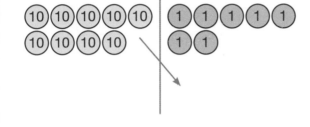

10s	1s

10s	1s

3 Laila sorts 89 newspapers into 9 equal piles. How many newspapers are in each pile? ⬚ How many are leftover? ⬚ Model the division using place value counters.

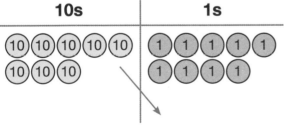

10s	1s

10s	1s

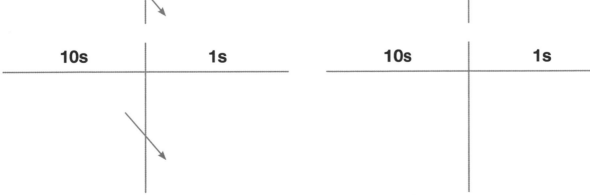

10s	1s

10s	1s

Date: _____

Number

Lesson 2: **Dividing 2-digit numbers by 1-digit numbers (2)**

- Use the expanded written method and short division to divide 2-digit numbers by 1-digit numbers

1 Estimate first. Then use the expanded written method of division to work out the answer to each calculation.

a 84 ÷ 3 = ☐

Estimate: ☐

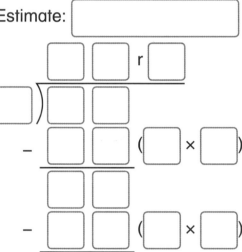

b 96 ÷ 4 = ☐

Estimate: ☐

c 87 ÷ 5 = ☐

Estimate: ☐

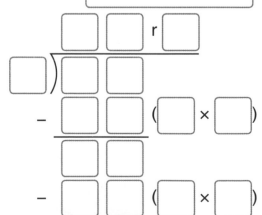

d 89 ÷ 6 = ☐

Estimate: ☐

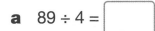

2 Estimate first, then use short division to work out the answer to each calculation.

a 89 ÷ 4 = ☐

Estimate: ☐

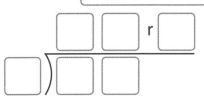

b 98 ÷ 5 = ☐

Estimate: ☐

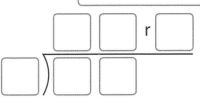

c 87 ÷ 7 = ☐

Estimate: ☐

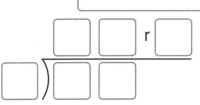

d 95 ÷ 4 = ☐

Estimate: ☐

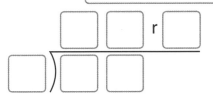

3 Solve the word problems. Use the expanded written method or short division to complete each calculation. Show your working in the box.

a Zane shares 97 g of sugar equally between 5 cups. How much sugar is in each cup? Give your answer as a mixed number.

Sugar in each cup: ☐ g

b A mouse travels 94 metres in 8 minutes. How far does it travel in 1 minute? Give your answer as a mixed number.

Total distance: ☐ m

Date: _____

Number

Lesson 3: **Dividing 3-digit numbers by 1-digit numbers (1)**

- Use the 'chunking' method to divide 3-digit numbers by 1-digit numbers

You will need

- paper for working out

1 Write the missing numbers and find the quotient. Example:

$224 \div 4 = (\boxed{200} \div 4) + (\boxed{24} \div \boxed{4}) = \boxed{50} + \boxed{6} = \boxed{56}$

a $177 \div 3 = (\boxed{} \div 3) + (\boxed{} \div 3) = \boxed{} + \boxed{} = \boxed{}$

b $384 \div 6 = (\boxed{} \div 6) + (\boxed{} \div 6) = \boxed{} + \boxed{} = \boxed{}$

c $308 \div 7 = (\boxed{} \div 7) + (\boxed{} \div 7) = \boxed{} + \boxed{} = \boxed{}$

d $512 \div 8 = (\boxed{} \div 8) + (\boxed{} \div 8) = \boxed{} + \boxed{} = \boxed{}$

2 Use place value counters to model each division calculation. Estimate the answer first. Complete the steps to find the quotient.

a $311 \div 4 = \boxed{}$ Estimate: $\boxed{}$

100s	10s	1s	100s	10s	1s
100 100 100	10	1			

b $265 \div 6 = \boxed{}$ Estimate: $\boxed{}$

100s	10s	1s	100s	10s	1s
100 100	10 10 10 10 10 10	1 1 1 1 1			

Number

c   254 ÷ 3 = ☐   Estimate: ☐

100s	10s	1s	100s	10s	1s

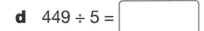

d   449 ÷ 5 = ☐   Estimate: ☐

100s	10s	1s	100s	10s	1s

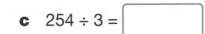

3   Solve the calculations using any method you prefer, mental or written. Use paper for working out if you need to. These divisions have no remainders.

a   296 ÷ 4 = ☐   **b**   291 ÷ 3 = ☐   **c**   312 ÷ 6 = ☐

d   182 ÷ 7 = ☐   **e**   495 ÷ 5 = ☐   **f**   368 ÷ 8 = ☐

These divisions have remainders.

g   547 ÷ 6 = ☐   **h**   785 ÷ 8 = ☐   **i**   266 ÷ 3 = ☐

j   253 ÷ 4 = ☐   **k**   436 ÷ 9 = ☐   **l**   523 ÷ 8 = ☐

4   Find five 3-digit numbers that give a remainder of 4 when divided by 6. Each 3-digit number must have a different hundreds digit. Write the numbers and describe the strategy you used.

_____

_____

Date: _____

Lesson 4: **Dividing 3-digit numbers by 1-digit numbers (2)**

- Use the expanded written method and short division to divide 3-digit numbers by 1-digit numbers

1 Estimate first, then use the expanded written method of division to work out the answer to each calculation.

a 276 ÷ 6 = ⬚

Estimate: ⬚

b 372 ÷ 4 = ⬚

Estimate: ⬚

c 263 ÷ 3 = ⬚

Estimate: ⬚

d 576 ÷ 7 = ⬚

Estimate: ⬚

2 Estimate first, then use short division to work out the answer to each calculation.

a   378 ÷ 5 =

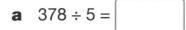

Estimate:

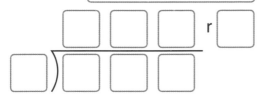

b   233 ÷ 4 =

Estimate:

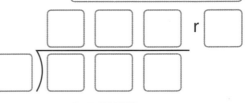

c   687 ÷ 8 =

Estimate:

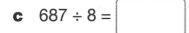

d   541 ÷ 7 =

Estimate:

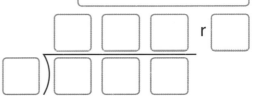

3 Solve the word problems. Use the expanded written method or short division to complete each calculation. Show your working in the box.

a $576 is shared equally between 8 people. How much does each person get?

$

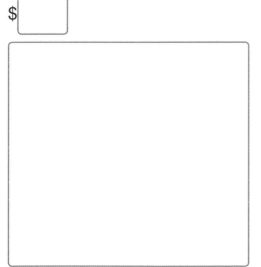

b   Water flows into a large tank at a rate of 469 litres every 9 minutes. How much water flows every minute? Write your answer as a mixed number.

Total: [ ] litres

Date: _____

Number

Lesson 1: **Dividing 2-digit numbers by 2-digit numbers (1)**

- Use the expanded written method to divide 2-digit numbers by 2-digit numbers

You will need

- paper for working out

1 Work out the multiplications mentally.

a 3 × 17 = ☐

b 4 × 23 = ☐

c 5 × 18 = ☐

d 6 × 13 = ☐

e 4 × 19 = ☐

f 3 × 26 = ☐

g 6 × 16 = ☐

h 4 × 24 = ☐

i 6 × 14 = ☐

j 3 × 23 = ☐

k 3 × 29 = ☐

l 5 × 17 = ☐

2 Use the expanded written method to solve each calculation. Estimate the answer first.

a 72 ÷ 18 = ☐

Estimate: ☐

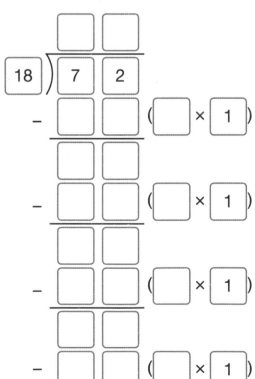

Number

b $78 \div 26 =$ ☐

Estimate: ☐

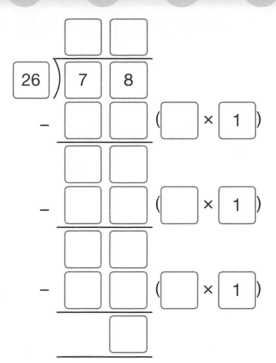

△**3** Use the expanded written method to solve each calculation.
Draw your own layout and set of steps on a separate piece of paper.

a $84 \div 28 =$ ☐

Estimate: ☐

b $96 \div 32 =$ ☐

Estimate: ☐

4 Solve the word problems on a separate piece of paper. Use the expanded written method of division.

a Nadiya buys 16 packets of football stickers. Each packet contains 6 stickers. She fixes the stickers in the pages of an album. If each page can hold 32 stickers, how many pages does she fill?

☐ pages

b Ekon cuts paper into strips of 14 cm. He attaches six strips together to make one long piece of paper. Ekon then cuts the long strip into equal sections of 28 cm.

How many strips does he now have? ☐ strips

Date: _____

Lesson 2: **Dividing 2-digit numbers by 2-digit numbers (2)**

Number

- Use the compact form of the expanded written method to divide 2-digit numbers by 2-digit numbers

1 Use the expanded written method to solve each calculation. Estimate the answer first.

a $76 \div 19 = \boxed{}$

Estimate: $\boxed{}$

$$\boxed{}\boxed{}$$
$$19 \overline{\smash{)} \ 7 \ \ 6 }$$
$$- \ \boxed{}\boxed{} \quad (\boxed{} \times \boxed{1})$$
$$\boxed{}\boxed{}$$
$$- \ \boxed{}\boxed{} \quad (\boxed{} \times \boxed{1})$$
$$\boxed{}\boxed{}$$
$$- \ \boxed{}\boxed{} \quad (\boxed{} \times \boxed{1})$$
$$\boxed{}\boxed{}$$
$$- \ \boxed{}\boxed{} \quad (\boxed{} \times \boxed{1})$$
$$\boxed{}$$

b $51 \div 17 = \boxed{}$

Estimate: $\boxed{}$

$$\boxed{}\boxed{}$$
$$17 \overline{\smash{)} \ 5 \ \ 1 }$$
$$- \ \boxed{}\boxed{} \quad (\boxed{} \times \boxed{1})$$
$$\boxed{}\boxed{}$$
$$- \ \boxed{}\boxed{} \quad (\boxed{} \times \boxed{1})$$
$$\boxed{}\boxed{}$$
$$- \ \boxed{}\boxed{} \quad (\boxed{} \times \boxed{1})$$
$$\boxed{}$$

2 Use the compact form of the expanded written method to solve each calculation. Estimate the answer first.

a $96 \div 12 = \boxed{}$

Estimate: $\boxed{}$

b $96 \div 32 = \boxed{}$

Estimate: $\boxed{}$

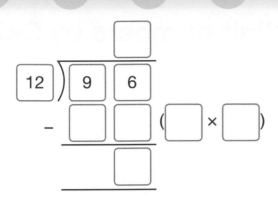
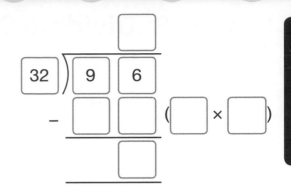

 3 Use the compact form of the expanded written method to solve each calculation. Draw your own layout and set of steps.

a   76 ÷ 38 =

Estimate:

b   76 ÷ 19 =

Estimate:

4   Solve the word problems. Use the compact form of the expanded written method of division.

a   A theatre has two areas of 36 seats. All the seats are occupied. If all the people arrive in groups of 12, how many groups is that?

    groups

b   13 out of 17 birds around a lake are flamingos. How many flamingos will there be in a lake with 68 birds?

    flamingos

Date: _____

Number

Lesson 3: **Dividing 3-digit numbers by 2-digit numbers (1)**

* Use the expanded written method to divide 3-digit numbers by 2-digit numbers

1 Draw a ring around the multiplication calculation that gives the best estimate.

a	414 ÷ 18	18 × 10	18 × 20	18 × 30	18 × 40
b	782 ÷ 23	23 × 10	23 × 20	23 × 30	23 × 40
c	704 ÷ 16	16 × 10	16 × 20	16 × 30	16 × 40
d	918 ÷ 27	27 × 10	27 × 20	27 × 30	27 × 40
e	576 ÷ 32	32 × 10	32 × 20	32 × 30	32 × 40
f	598 ÷ 46	46 × 10	46 × 20	46 × 30	46 × 40

2 Use the expanded written method to solve each calculation. Estimate the answer first.

a 468 ÷ 26 = ☐

Estimate: ☐

b 988 ÷ 38 = ☐

Estimate: ☐

Number

3 Use the expanded written method to solve each calculation. Draw your own layout and set of steps.

a 598 ÷ 23 = ☐

Estimate: ☐

b 612 ÷ 34 = ☐

Estimate: ☐

4 Use the long division method to solve each word problem.

a Grace buys several computer games for a total of $648. If each game costs $27, how many games does Grace buy?

☐ games

b A café makes 684 baklava pastries to be delivered to a party. If they pack 36 in a box, how many boxes will they require to pack all the pastries?

☐ boxes

Date: _____

Lesson 4: **Dividing 3-digit numbers by 2-digit numbers (2)**

 Number

• Use the long division method to divide 3-digit numbers by 2-digit numbers

1 Use the expanded written method to solve each calculation. Estimate the answer first.

a 816 ÷ 24 = ☐

Estimate: ☐

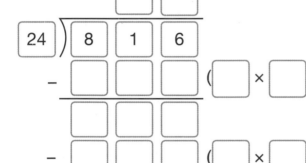

b 848 ÷ 16 = ☐

Estimate: ☐

2 Use the long division method to solve each calculation. Estimate the answer first.

a 828 ÷ 18 = ☐

Estimate: ☐

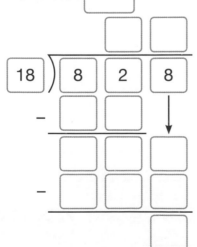

b 936 ÷ 36 = ☐

Estimate: ☐

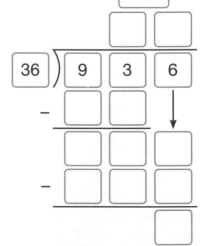

Number

3 Use the long division method to solve each calculation. Draw your own layout and set of steps.

a 675 ÷ 27 = [ ]

Estimate: [ ]

b 784 ÷ 14 = [ ]

Estimate: [ ]

4 Use the long division method to solve each word problem.

a 6 out of every 19 shapes are triangles. How many triangles will there be in a set of 513 shapes?

[ ] triangles

b 19 out of every 34 cars are red. How many cars will be red in a group of 918 cars?

[ ] cars

Date: _____

Number

Lesson 1: **Decimal place value**

- Understand and explain the value of each digit in decimals

1 In each box, write the number the arrow points to.

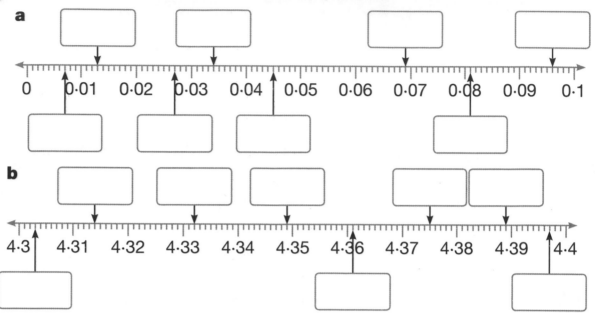

2 Write the value of each underlined digit.

Example: 2·46<u>7</u> [*7 thousandths or 0·007*]

a 0·<u>5</u>86 [ ]   **b** 2·4<u>6</u>8 [ ]

c 9·23<u>1</u> [ ]   **d** −0·1<u>1</u>1 [ ]

e 27·<u>9</u>99 [ ]   **f** 0·44<u>4</u> [ ]

g 5·2<u>0</u>6 [ ]   **h** −37·86<u>2</u> [ ]

3 Write the decimal that is equivalent to each fraction.

a $\frac{3}{1000}$ = [ ]   **b** $\frac{7}{1000}$ = [ ]

c $\frac{19}{1000}$ = [ ]   **d** $\frac{68}{1000}$ = [ ]

e $\frac{484}{1000}$ = [ ]   **f** $\frac{987}{1000}$ = [ ]

Number

4 Write the next three numbers in each sequence.

Counting on in steps of…						
0·001	0·057	0·058	0·059			
0·002	0·293	0·295	0·297			
0·003	−0·526	−0·523	−0·52			
0·005	−3·287	−3·282	−3·277			

Counting back in steps of…						
0·001	2·474	2·473	2·472			
0·002	0·007	0·005	0·003			
0·003	−0·535	−0·537	−0·539			
0·005	−6·748	−6·753	−6·758			

5 Specialise and write an example of a number with:

a 8 thousandths between 1·2 and 1·3

b 6 hundredths between 4·8 and 4·9

c 37 thousandths

d negative 8 thousandths between −6·2 and −6·3

e negative 5 thousandths between −4·18 and −4·19

f negative 99 thousandths

Date: _____

71

Number

Lesson 2: **Composing and decomposing decimals**

• Compose and decompose decimals

 Draw lines to match the decimals to their decomposed forms.

3 + 0·7 + 0·09 + 0·003	3 ones + 9 tenths + 7 thousandths

3·907   3·709   3·097   3·793

3 ones + 9 hundredths + 7 thousandths   3 + 0·7 + 0·009

2 Decompose the numbers by place value.

a 0·526 = ☐ + ☐ + ☐

b 0·928 = ☐ + ☐ + ☐

c 0·303 = ☐ + ☐

d 0·066 = ☐ + ☐

e 5·555 = ☐ + ☐ + ☐ + ☐

f 8·008 = ☐ + ☐

 Complete the sentences.

5·837 is composed from 5 + 0·8 + 0·03 + 0·007.

a 1·234 is composed from ☐

b 6·019 is composed from ☐

c 11·746 is composed from ☐

d   54·802 is composed from

e   476·333 is composed from

f   777·777 is composed from

g   432·057 is composed from

4   Find four different ways to decompose each number.

Example: 2·486

i   2 + 0·4 + 0·08 + 0·006   **ii**   24 tenths and 86 thousandths

iii   248 hundredths and 6 thousandths   **iv**   2 ones and 486 thousandths

a   3·294

i   **ii**

iii   **iv**

b   5·171

i   **ii**

iii   **iv**

c   9·658

i   **ii**

iii   **iv**

d   6·039

i   **ii**

iii   **iv**

Date: _____

Number

Lesson 3: **Regrouping decimals**

- Regroup decimals to help with calculations

1 Decompose each number by place value.

Example: 32·179 = 30 + 2 + 0.1 + 0.07 + 0·009

a 2·281 = ☐ + ☐ + ☐ + ☐

b 8·432 = ☐ + ☐ + ☐ + ☐

c 27·437 = ☐ + ☐ + ☐ + ☐

+ ☐

d 63·056 = ☐ + ☐ + ☐ + ☐

e 81·208 = ☐ + ☐ + ☐ + ☐

f 548·453 = ☐ + ☐ + ☐ + ☐

+ ☐ + ☐

2 Write each decimal.

a 4 ones, 2 tenths, 6 hundredths and 5 thousandths ☐

b 58 tenths and 19 thousandths ☐

c 40 tens, 4 ones and 187 thousandths ☐

d 708 hundredths and 2 thousandths ☐

3 Decompose each number in four different ways

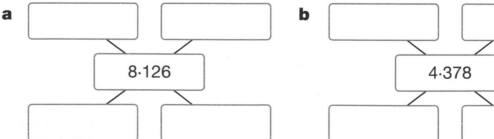

a ☐ ☐

8·126

☐ ☐

b ☐ ☐

4·378

☐ ☐

Number

c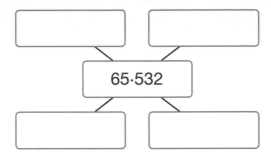

65·532

d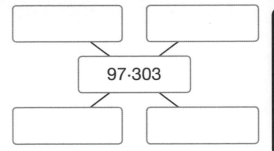

97·303

4 **a** Use the number line to help you regroup each number.

i −26 =

ii −51 =

b Use the number line to help you decompose each number.

i −4·173 =

ii −13·284 =

5 How would you use decomposition and regrouping to solve each subtraction problem?

Write your working in the box provided.

a 8·363 − 4·138 =

b 43·746 − 21·824 =

(Clue: Think 8·3 − 4·1 + 0·063 − 0·038)

Date: _____

Number

Lesson 4: Comparing and ordering decimals

- Compare and order decimals with two decimal places

1 Compare the decimals using the symbols < or >.

a 2·34 ☐ 2·43

10s	1s	0.1s	0.01s
	2 •	3	4
	2 •	4	3

b 5·21 ☐ 5·23

10s	1s	0.1s	0.01s
	5 •	2	1
	5 •	2	3

c 17·03 ☐ 17·02

10s	1s	0.1s	0.01s
1	7 •	0	3
1	7 •	0	2

d 65·46 ☐ 64·56

10s	1s	0.1s	0.01s
6	5 •	4	6
6	4 •	5	6

2 Compare the decimals using the symbols < or >.

a 0·61 ☐ 0·16 **b** 0·98 ☐ 0·89 **c** 2·67 ☐ 2·76

d 5·43 ☐ 5·41 **e** 8·17 ☐ 8·71 **f** 6·04 ☐ 6·03

g 12·23 ☐ 12·32 **h** 15·26 ☐ 12·56 **i** 22·21 ☐ 21·22

j 37·73 ☐ 37·37 **k** 45·04 ☐ 45·03 **l** 82·23 ☐ 82·32

3 Use the place value charts to help you order the numbers.

a

10s	1s	0.1s	0.01s
	•		
	•		
	•		
	•		
	•		

6·42, 6·37, 6·36, 6·24, 6·41

☐ < ☐ < ☐ < ☐ < ☐

Number

b

10s	1s	0.1s	0.01s
		•	
		•	
		•	
		•	
		•	

32·76, 32·86, 32·68, 31·68, 32·67

c

10s	1s	0.1s	0.01s
		•	
		•	
		•	
		•	
		•	

4·13, 4, 4·31, 4·11, 4·1

d

10s	1s	0.1s	0.01s
		•	
		•	
		•	
		•	
		•	

56·56, 55·66, 56·5, 56·65, 56

4 Fill in the missing digits to make each statement true.

a 2·4 < 2· ☐

b 210·5 ☐ > 210· ☐ 9

c 13· ☐ 4 > 13· ☐ 9

d 199· ☐ 9 < 199· ☐ 9

e 20·8 ☐ < 20·8 ☐

f 105·1 ☐ > 105· ☐ 8

g 10·0 ☐ > 1 ☐ · ☐ 3

h 400·0 ☐ < 400·0 ☐

i 37· ☐ < 37· ☐ 5

j 354· ☐ 6 < 354.5 ☐

Date: _____

77

Number

Lesson 1: **Multiplying whole numbers and decimals by 10, 100 and 1000**

• Multiply whole numbers and decimals by 10, 100 and 1000

1 Use the place value charts to complete the calculations.

a 28 × 100 = [ ]

100000s	10000s	1000s	100s	10s	1s	0·1s	0·01s
					•		
					•		

b 327 × 1000 = [ ]

100000s	10000s	1000s	100s	10s	1s	0·1s	0·01s
					•		
					•		

c 7·9 × 100 = [ ]

100000s	10000s	1000s	100s	10s	1s	0·1s	0·01s
					•		
					•		

d 45·68 × 1000 = [ ]

100000s	10000s	1000s	100s	10s	1s	0·1s	0·01s
					•		
					•		

2 Write the missing numbers in the table.

Number	× 10	× 100	× 1000
25		2500	
98	980		
303			303 000
2·9	29		
8·06		806	
55·55	555·5		

Number

3 Complete the missing numbers in the calculations with 10, 100 or 1000.

a 8·7 × ☐ = 870

b 13·4 × ☐ = 134

c 4·56 × ☐ = 4560

d 6·06 × ☐ = 606

e 19·8 × ☐ = 19 800

f 70·2 × ☐ = 702

4 Fill in the outputs.

a

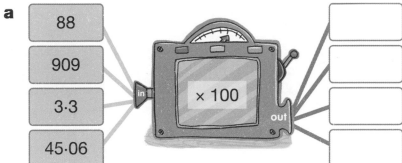

88
909
3·3
45·06

in × 100 out

b

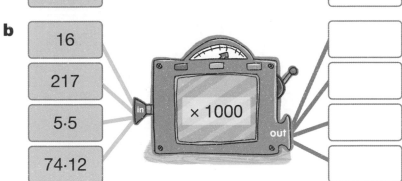

16
217
5·5
74·12

in × 1000 out

5 Fill in the missing digits.

a ☐ 5· ☐ 7 × 100 = 8547

b 9 ☐ 2·3 ☐ × 10 = 912 ☐ ·8

c ☐ 43· ☐ 8 × ☐ = 7 ☐ 3 080

d 7 ☐ ☐ ·15 × 100 = 734 ☐ ☐

e 546 ☐ ·3 ☐ × ☐ = 5 ☐ 65 ☐ 20

f ☐ 94 ☐ ·09 × ☐ = 3 ☐ 420· ☐

Number

Lesson 2: **Dividing whole numbers and decimals by 10, 100 and 1000**

• Divide whole numbers and decimals by 10, 100 and 1000

1 Use the place value charts to complete the calculations.

a 4820 ÷ 10 = ☐

1000s	100s	10s	1s	0·1s	0·01s	0·001s
			•			
			•			

b 5870 ÷ 1000 = ☐

1000s	100s	10s	1s	0·1s	0·01s	0·001s
			•			
			•			

c 645·1 ÷ 100 = ☐

1000s	100s	10s	1s	0·1s	0·01s	0·001s
			•			
			•			

d 76069 ÷ 1000 = ☐

10000s	1000s	100s	10s	1s	0·1s	0·01s	0·001s
				•			
				•			

2 Write the missing numbers in the table.

Number	÷ 10	÷ 100	÷ 1000
99900		999	
2121·6			▓▓▓
46775	4677·5		
101		1·01	
56·14		▓▓▓	▓▓▓
889865			889·865

Number

3 Complete the missing numbers in the calculations with 10, 100 or 1000.

a 235 ÷ ☐ = 2·35   **b** 48·4 ÷ ☐ = 4·84

c 8·9 ÷ ☐ = 0·89   **d** 88 700 ÷ ☐ = 88·7

e 103 ÷ ☐ = 1·03   **f** 50 550 ÷ ☐ = 50·55

4 Fill in the outputs.

a

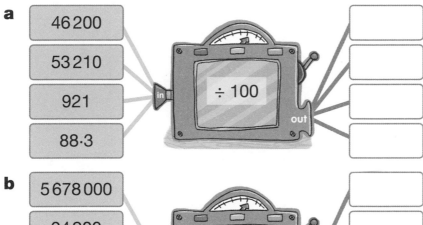

46 200
53 210
921
88·3

÷ 100

b

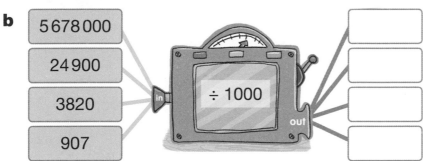

5 678 000
24 900
3820
907

÷ 1000

5 Fill in the missing digits.

a ☐ 123 ÷ 100 = 41·2 ☐

b ☐ 193 ÷ 10 = 819· ☐

c 3 ☐ 85 ÷ ☐ = 3·48 ☐

d 2470 ☐ ☐ ÷ ☐ = 2470

e 7· ☐ 3 ÷ 10 = 0· ☐ 6 ☐

f 8 ☐ 5 ÷ 1000 = 0· ☐ 0 ☐

Date: _____

Number

Lesson 3: **Rounding decimals to the nearest tenth**

• Round decimals to the nearest tenth

1 Round to the nearest tenth.

a

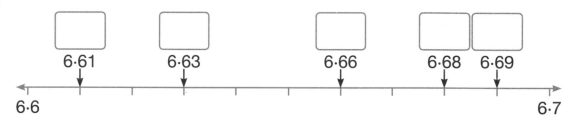

b

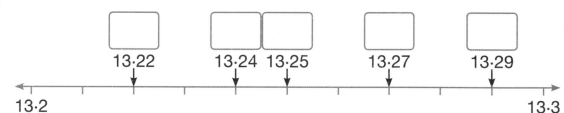

2 Write the two one-place decimals that each number comes between. Then draw a ring around the number that the decimal rounds to. The first one has been done for you.

a (1·3) | 1·34 | 1·4   **b** [ ] 2·86 [ ]

c [ ] 8·15 [ ]   **d** [ ] 9·61 [ ]

e [ ] 17·98 [ ]   **f** [ ] 43·67 [ ]

g [ ] 74·22 [ ]   **h** [ ] 96·55 [ ]

3 Round each measurement to the nearest tenth.

a 8·77 kg [ ] kg   **b** 7·35 m [ ] m

c 3·63 *l* [ ] *l*   **d** 5·97 km [ ] km

e 12·84 g [ ] g   **f** 19·22 cm [ ] cm

Number

g 23·48 seconds ☐ seconds

h 50·01 m ☐ m   **i** 89·93 kg ☐ kg

j 92·89 *l* ☐ *l*   **k** 65·84 km ☐ km

4 Round the prices to the nearest tenth of a dollar.

Remember to include a zero digit in the hundredths position.

Example: $3.62 *$3.60*

a $2.34 ☐   **b** $5.17 ☐

c $12.59 ☐   **d** $37.99 ☐

e $42.03 ☐   **f** $65.72 ☐

g $234.28 ☐   **h** $546.92 ☐

5 Write all the numbers with two decimal places that round to:

a 8·3 _____

b 14·7 _____

c 35·5 _____

d 86·1 _____

6 Write four positive numbers with two decimal places that:

a when rounded to the nearest tenth are zero.

☐ ☐ ☐ ☐

b when rounded to the nearest tenth are 1.

☐ ☐ ☐ ☐

Date: _____

Lesson 4: **Rounding decimals to the nearest whole number**

* Round decimals to the nearest whole number

1 Write the two whole numbers that each decimal comes between. Then draw a ring around the number that the decimal rounds to. The first one has been done for you.

a | 4 | 4·73 | (5)   **b** | | 6·35 | |

c | | 8·18 | | **d** | | 9·09 | |

e | | 11·76 | | **f** | | 16·92 | |

g | | 20·11 | | **h** | | 46·45 | |

2 Round each decimal to the nearest whole number.

Number	6·28	12·35	15·91	40·27	69·94
Rounded to the nearest whole number					

3 Round each measurement to the nearest whole number and the nearest tenth.

Number	16·37 cm	28·45 km	47·18 kg	52·99 l	$735.08
Rounded to the nearest whole number					
Rounded to the nearest tenth					

Number

4 Draw a ring around the mystery number described.

19·07   18·72   18·98   20·74   18·76   18·73

a I am between 10 and 20.

I am 19 rounded to the nearest whole number.

I am 18·7 rounded to the nearest tenth.

My hundredths digit is even.

81·12   90·13   80·13   81·17   90·11   81·11

b I am between 80 and 90.

I am 81 rounded to the nearest whole number.

I am 81·1 rounded to the nearest tenth.

My hundredths digit is odd.

233·43   233·53   241·43   233·78   235·54   233·75

c I am between 231·98 and 241·34

I am 234 rounded to the nearest whole number.

I am 233·8 rounded to the nearest tenth.

My hundredths digit is even.

5 Write the missing numbers in the table. There is more than one answer.

Number					
Rounded to the nearest whole number					
Rounded to the nearest tenth	12·4	17·9	27·4	59·1	92·6

Date: _____

Lesson 1: **Fractions as division**

 Number

• Understand that a fraction can be thought of as a division of the numerator by the denominator

1 Write each fraction as a division.

a $\frac{1}{4}$ = ☐ ÷ ☐   **b** $\frac{1}{3}$ = ☐ ÷ ☐   **c** $\frac{1}{5}$ = ☐ ÷ ☐

d $\frac{3}{5}$ = ☐ ÷ ☐   **e** $\frac{1}{8}$ = ☐ ÷ ☐   **f** $\frac{5}{8}$ = ☐ ÷ ☐

g $\frac{1}{16}$ = ☐ ÷ ☐   **h** $\frac{11}{16}$ = ☐ ÷ ☐   **i** $\frac{5}{4}$ = ☐ ÷ ☐

j $\frac{8}{7}$ = ☐ ÷ ☐   **k** $\frac{19}{17}$ = ☐ ÷ ☐   **l** $\frac{23}{20}$ = ☐ ÷ ☐

2 Write each division as a fraction.

a $2 \div 3 = \dfrac{☐}{☐}$ **b** $4 \div 5 = \dfrac{☐}{☐}$ **c** $7 \div 8 = \dfrac{☐}{☐}$

d $3 \div 10 = \dfrac{☐}{☐}$ **e** $11 \div 15 = \dfrac{☐}{☐}$ **f** $7 \div 20 = \dfrac{☐}{☐}$

g $6 \div 5 = \dfrac{☐}{☐}$ **h** $14 \div 11 = \dfrac{☐}{☐}$ **i** $19 \div 16 = \dfrac{☐}{☐}$

3 Write the fraction described and then express the fraction as a division.

a There are 4 cars in a car park. 3 of the cars are green.

Number of green cars as a fraction of the total number of cars: $\dfrac{☐}{☐}$

Fraction as a division: ☐ ÷ ☐

b There are 8 pieces of fruit in a bowl – 3 of them are bananas.

Number of bananas as a fraction of the total number of fruit:

Fraction as a division: ▢ ÷ ▢

c There are 20 shapes in the tray. 7 out of the shapes are squares.

Number of squares as a fraction of the total number of shapes:

Fraction as a division: ▢ ÷ ▢

d There are 9 girls in a group of 20 children.

Number of girls as a fraction of the total number of children:

Fraction as a division: ▢ ÷ ▢

4 Convert each mixed number to an improper fraction and then write the fraction as a division.

a $1\frac{1}{4} = \dfrac{\square}{\square} = \square ÷ \square$ **b** $1\frac{2}{3} = \dfrac{\square}{\square} = \square ÷ \square$

c $2\frac{3}{4} = \dfrac{\square}{\square} = \square ÷ \square$ **d** $1\frac{3}{5} = \dfrac{\square}{\square} = \square ÷ \square$

e $2\frac{4}{5} = \dfrac{\square}{\square} = \square ÷ \square$ **f** $3\frac{1}{8} = \dfrac{\square}{\square} = \square ÷ \square$

g $4\frac{3}{8} = \dfrac{\square}{\square} = \square ÷ \square$ **h** $5\frac{7}{10} = \dfrac{\square}{\square} = \square ÷ \square$

Date: _____

Lesson 2: **Simplifying fractions**

• Simplify a fraction to its lowest terms

1 Use the Venn diagrams to record the factors of each number. Remember to put common factors in the overlap. Draw a ring around the highest common factor.

a Reduce $\frac{24}{30}$ to its simplest form.

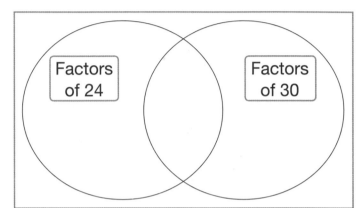

b Reduce $\frac{18}{45}$ to its simplest form.

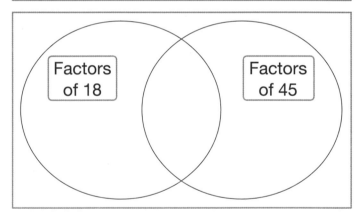

2 Simplify each fraction in one or two steps.

Example 1:

$$\frac{15}{25} = \boxed{\frac{15}{25} = \frac{3}{5}} = \boxed{\frac{3}{5}}$$

Example 2:

$$\frac{16}{32} = \boxed{\frac{16}{32} = \frac{8}{16} = \frac{1}{2}} = \boxed{\frac{1}{2}}$$

a $\frac{9}{36} = $ ⬚ = ⬚

b $\frac{45}{60} = $ ⬚ = ⬚

Number

c $\frac{24}{60}$ = ☐ = ☐   **d** $\frac{49}{70}$ = ☐ = ☐

e $\frac{72}{80}$ = ☐ = ☐   **f** $\frac{68}{84}$ = ☐ = ☐

3 Draw lines to match each fraction to its simplest form.

$\frac{36}{45}$ $\frac{48}{96}$ $\frac{12}{48}$ $\frac{56}{80}$ $\frac{30}{75}$ $\frac{27}{36}$

$\frac{3}{4}$ $\frac{2}{5}$ $\frac{7}{10}$ $\frac{1}{4}$ $\frac{4}{5}$ $\frac{1}{2}$

4 Write the fractions with the same denominator, then solve the addition.
Write the answer as a mixed number. The first one has been done for you.

a $\frac{1}{5} + \frac{1}{2} + \frac{4}{10}$ = $\frac{2}{10}$ + $\frac{5}{10}$ + $\frac{4}{10}$ = $\frac{11}{10}$ = $1\frac{1}{10}$

b $\frac{2}{5} + \frac{1}{2} + \frac{6}{10}$ = $\frac{\square}{\square}$ + $\frac{\square}{\square}$ + $\frac{\square}{\square}$ = $\frac{\square}{\square}$ = ☐

c $\frac{3}{4} + \frac{3}{8} + \frac{1}{8}$ = $\frac{\square}{\square}$ + $\frac{\square}{\square}$ + $\frac{\square}{\square}$ = $\frac{\square}{\square}$ = ☐

d $\frac{9}{10} + \frac{4}{5} + \frac{1}{2}$ = $\frac{\square}{\square}$ + $\frac{\square}{\square}$ + $\frac{\square}{\square}$ = $\frac{\square}{\square}$ = ☐

Date: _____

Lesson 3: **Comparing fractions with different denominators**

- Compare fractions with different denominators

1 Draw a ring around the smaller fraction in each pair. Use the fraction wall to help you.

1								
$\frac{1}{3}$			$\frac{1}{3}$			$\frac{1}{3}$		
$\frac{1}{6}$		$\frac{1}{6}$	$\frac{1}{6}$		$\frac{1}{6}$	$\frac{1}{6}$		$\frac{1}{6}$
$\frac{1}{9}$	$\frac{1}{9}$	$\frac{1}{9}$	$\frac{1}{9}$	$\frac{1}{9}$	$\frac{1}{9}$	$\frac{1}{9}$	$\frac{1}{9}$	$\frac{1}{9}$

a $\frac{1}{6}$ $\frac{1}{3}$ **b** $\frac{1}{6}$ $\frac{1}{9}$ **c** $\frac{3}{6}$ $\frac{1}{3}$

d $\frac{2}{9}$ $\frac{1}{6}$ **e** $\frac{2}{3}$ $\frac{4}{9}$ **f** $\frac{4}{6}$ $\frac{1}{3}$

g $\frac{5}{9}$ $\frac{2}{3}$ **h** $\frac{3}{3}$ $\frac{5}{6}$ **i** $\frac{5}{6}$ $\frac{8}{9}$

2 Use the diagram to convert one fraction to have the same denominator as the other. Then compare the fractions using < or >. The first one has been done for you.

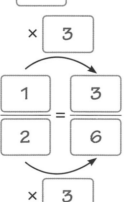

a $\frac{1}{2}$ < $\frac{5}{6}$

× 3

$\frac{1}{2} = \frac{3}{6}$

× 3

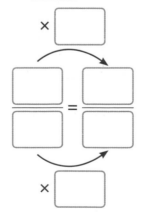

b $\frac{7}{8}$ ☐ $\frac{3}{4}$

× ☐

☐/☐ = ☐/☐

× ☐

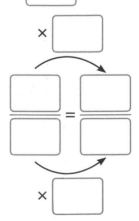

c $\frac{5}{8}$ ☐ $\frac{11}{16}$

× ☐

☐/☐ = ☐/☐

× ☐

d $\frac{3}{8}$ ☐ $\frac{1}{2}$

× ☐

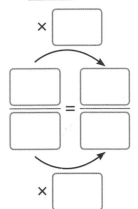

☐/☐ = ☐/☐

× ☐

e $\frac{4}{5}$ ☐ $\frac{7}{10}$

× ☐

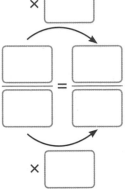

☐/☐ = ☐/☐

× ☐

f $\frac{2}{3}$ ☐ $\frac{9}{15}$

× ☐

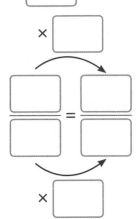

☐/☐ = ☐/☐

× ☐

3 Compare fractions using < or >.

a $\frac{1}{2}$ ☐ $\frac{3}{4}$

b $\frac{2}{10}$ ☐ $\frac{2}{5}$

c $\frac{3}{4}$ ☐ $\frac{5}{8}$

d $\frac{2}{3}$ ☐ $\frac{7}{9}$

e $\frac{2}{5}$ ☐ $\frac{3}{10}$

f $\frac{2}{3}$ ☐ $\frac{5}{6}$

4 Compare the fractions by converting them to decimals. The first one has been done for you.

a $\frac{1}{2}$ or $\frac{1}{4}$ $\boxed{0.5 > 0.25}$

b $\frac{3}{10}$ or $\frac{1}{5}$ ☐

c $\frac{3}{4}$ or $\frac{8}{10}$ ☐

d $\frac{3}{5}$ or $\frac{7}{10}$ ☐

e $\frac{2}{5}$ or $\frac{3}{10}$ ☐

f $\frac{3}{5}$ or $\frac{1}{2}$ ☐

5 Compare numbers using < or >. Write your working in the box provided.

a $\frac{3}{5}$ ☐ 0.5

☐

b 0.8 ☐ $\frac{7}{10}$

☐

c $\frac{4}{5}$ ☐ 0.9

☐

d 0.75 ☐ $\frac{8}{10}$

☐

e $\frac{2}{5}$ ☐ 0.25

☐

f $\frac{6}{10}$ ☐ 0.5

☐

Date: _____

Lesson 4: **Ordering fractions with different denominators**

Number

• Order fractions with different denominators

1 Order each set of fractions, from least to greatest. Use the fraction wall to help you.

1								
$\frac{1}{3}$			$\frac{1}{3}$			$\frac{1}{3}$		
$\frac{1}{6}$		$\frac{1}{6}$	$\frac{1}{6}$		$\frac{1}{6}$	$\frac{1}{6}$		$\frac{1}{6}$
$\frac{1}{9}$	$\frac{1}{9}$	$\frac{1}{9}$	$\frac{1}{9}$	$\frac{1}{9}$	$\frac{1}{9}$	$\frac{1}{9}$	$\frac{1}{9}$	$\frac{1}{9}$

a $\frac{1}{3}, \frac{2}{3}, \frac{1}{9}$

Order: ☐☐ < ☐☐ < ☐☐

b $\frac{3}{6}, \frac{2}{3}, \frac{2}{6}$

Order: ☐☐ < ☐☐ < ☐☐

c $\frac{5}{9}, \frac{2}{3}, \frac{4}{9}$

Order: ☐☐ < ☐☐ < ☐☐

d $\frac{5}{6}, \frac{2}{3}, \frac{2}{6}$

Order: ☐☐ < ☐☐ < ☐☐

e $\frac{8}{9}, \frac{2}{3}, \frac{7}{9}$

Order: ☐☐ < ☐☐ < ☐☐

f $\frac{4}{6}, \frac{1}{6}, \frac{1}{3}$

Order: ☐☐ < ☐☐ < ☐☐

2 Convert the fractions to the same denominator and then order them.

a $\frac{1}{2}$ ☐ $\frac{6}{8}$ ☐ $\frac{1}{4}$ ☐

☐ < ☐ < ☐

b $\frac{2}{3}$ ☐ $\frac{4}{9}$ ☐ $\frac{1}{3}$ ☐

☐ < ☐ < ☐

Number

c $\frac{9}{10}$ $\frac{7}{10}$ $\frac{4}{5}$

☐ < ☐ < ☐

d $\frac{3}{4}$ $\frac{1}{2}$ $\frac{5}{8}$

☐ < ☐ < ☐

3 Convert the fractions to the same denominator and then order them.

a $\frac{3}{4}$ $\frac{3}{8}$ $\frac{1}{4}$ $\frac{5}{8}$

☐ < ☐ < ☐ < ☐

b $\frac{3}{5}$ $\frac{5}{10}$ $\frac{2}{5}$ $\frac{3}{10}$

☐ < ☐ < ☐ < ☐

c $\frac{2}{3}$ $\frac{5}{6}$ $\frac{1}{2}$ $\frac{3}{4}$

☐ < ☐ < ☐ < ☐

d $\frac{4}{9}$ $\frac{2}{6}$ $\frac{2}{3}$ $\frac{7}{9}$

☐ < ☐ < ☐ < ☐

4 Write the numerator and/or denominator to maintain the order.

a $\frac{1}{4} < \frac{1}{\square} < \frac{5}{8}$ **b** $\frac{1}{3} < \frac{\square}{\square} < \frac{5}{9}$ **c** $\frac{6}{10} < \frac{\square}{10} < \frac{4}{5}$

Date: _____

93

Number

Lesson 1: **Fractions as operators**

- Use a proper or improper fraction as an operator to find the fraction of a quantity

You will need
- paper for working out

1 Work out these unit fractions. Show your working out on a separate piece of paper.

a $\frac{1}{10}$ of \$90 = ☐   b $\frac{1}{5}$ of 25 km = ☐   c $\frac{1}{3}$ of 60 g = ☐

d $\frac{1}{4}$ of 120 m = ☐   e $\frac{1}{6}$ of 360 ml = ☐   f $\frac{1}{9}$ of 270 kg = ☐

g $\frac{1}{12}$ of \$720 = ☐   h $\frac{1}{15}$ of 300l = ☐   i $\frac{1}{25}$ of 450 cm = ☐

2 Work out these fractions. Show your working out on a separate piece of paper.

a $\frac{7}{10}$ of \$30 = ☐   b $\frac{3}{10}$ of 130 km = ☐   c $\frac{9}{10}$ of 310 g = ☐

d $\frac{3}{4}$ of 180 = ☐   e $\frac{2}{5}$ of \$600 = ☐   f $\frac{5}{6}$ of 300 km = ☐

g $\frac{3}{8}$ of 480 = ☐   h $\frac{7}{8}$ of \$608 = ☐   i $\frac{5}{7}$ of 581 km = ☐

j $\frac{4}{9}$ of 684 = ☐   k $\frac{7}{9}$ of \$792 = ☐   l $\frac{4}{11}$ of 726 km = ☐

3 Work out these fractions. Show your working out on a separate piece of paper.

a $\frac{4}{3}$ of \$27 = ☐   b $\frac{7}{3}$ of 69 km = ☐   c $\frac{10}{3}$ of 111 g = ☐

d $\frac{5}{4}$ of 64 = ☐   e $\frac{7}{4}$ of \$92 = ☐   f $\frac{9}{4}$ of 212 km = ☐

g $\frac{7}{5}$ of 85 = ☐   h $\frac{8}{5}$ of \$225 = ☐   i $\frac{11}{5}$ of 325 km = ☐

j $\frac{9}{8}$ of 128 = ☐   k $\frac{11}{8}$ of \$264 = ☐   l $\frac{13}{8}$ of 368 km = ☐

Number

4 Bella's teacher asks her to solve the problem:
$\frac{7}{5}$ of $75 =

Bella works out the answer like this:

Show Bella two other methods for solving this calculation.

> $\frac{1}{5}$ of $75 = 75 \div 5 = 15$
> $\frac{7}{5}$ of $75 = 7 \times 15$
> $\qquad = 105

Method 2	Method 3

5 Solve the word problems. Show your working out.

a Sabrina and Atique each make a jug of fruit drink. Sabrina uses $\frac{13}{5}$ times as much lemonade as Atique. Atique uses 165 ml of lemonade. What amount of lemonade does Sabrina use?

> The amount of lemonade used by Sabrina is ⬚ ml.

b On Monday, Mrs Jones drove 276 kilometres. On Tuesday, she drove $\frac{14}{6}$ times the distance she drove on Monday. How many kilometres did Mrs Jones drive on Tuesday?

> Distance driven on Tuesday: ⬚ km.

Date: _____

Number

Lesson 2: **Adding and subtracting fractions**

• Add and subtract fractions with different denominators

Estimate answers where possible. Write your estimates on a separate piece of paper.

Example: $\frac{4}{5} - \frac{1}{2}$ $\frac{4}{5}$ is close to 1 so the answer should be a bit less than $\frac{1}{2}$.

1 Solve the fraction calculations.

a $\frac{1}{5} + \frac{2}{5} = \boxed{}$

b $\frac{3}{6} + \frac{2}{6} = \boxed{}$

c $\frac{4}{5} - \frac{2}{5} = \boxed{}$

d $\frac{5}{6} - \frac{4}{6} = \boxed{}$

e $\frac{4}{8} + \frac{3}{8} = \boxed{}$

f $\frac{7}{8} - \frac{2}{8} = \boxed{}$

g $\frac{13}{10} - \frac{2}{10} = \boxed{}$

h $\frac{3}{4} + \frac{2}{4} = \boxed{}$

i $\frac{8}{7} + \frac{5}{7} = \boxed{}$

j $\frac{17}{12} - \frac{4}{12} = \boxed{}$

k $\frac{13}{10} + \frac{14}{10} = \boxed{}$

l $\frac{15}{9} - \frac{4}{9} = \boxed{}$

2 Solve the fraction calculations. Write answers greater than 1 as an improper fraction and a mixed number.
Example:

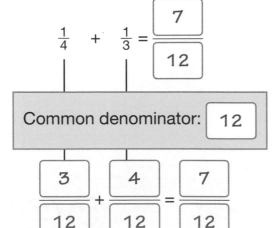

$$\frac{1}{4} + \frac{1}{3} = \frac{7}{12}$$

Common denominator: 12

$$\frac{3}{12} + \frac{4}{12} = \frac{7}{12}$$

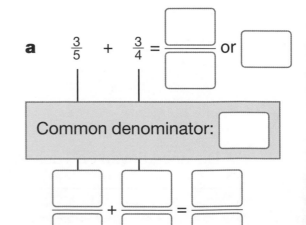

a $\frac{3}{5} + \frac{3}{4} = \frac{\boxed{}}{\boxed{}}$ or $\boxed{}$

Common denominator: $\boxed{}$

$$\frac{\boxed{}}{\boxed{}} + \frac{\boxed{}}{\boxed{}} = \frac{\boxed{}}{\boxed{}}$$

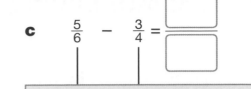

Number

b $\dfrac{4}{5} - \dfrac{1}{2} = \dfrac{\boxed{}}{\boxed{}}$

Common denominator: $\boxed{}$

$\dfrac{\boxed{}}{\boxed{}} - \dfrac{\boxed{}}{\boxed{}} = \dfrac{\boxed{}}{\boxed{}}$

c $\dfrac{5}{6} - \dfrac{3}{4} = \dfrac{\boxed{}}{\boxed{}}$

Common denominator: $\boxed{}$

$\dfrac{\boxed{}}{\boxed{}} - \dfrac{\boxed{}}{\boxed{}} = \dfrac{\boxed{}}{\boxed{}}$

d $\dfrac{5}{6} + \dfrac{4}{5} = \dfrac{\boxed{}}{\boxed{}}$ or $\boxed{}$

Common denominator: $\boxed{}$

$\dfrac{\boxed{}}{\boxed{}} + \dfrac{\boxed{}}{\boxed{}} = \dfrac{\boxed{}}{\boxed{}}$

e $\dfrac{9}{10} - \dfrac{5}{9} = \dfrac{\boxed{}}{\boxed{}}$

Common denominator: $\boxed{}$

$\dfrac{\boxed{}}{\boxed{}} - \dfrac{\boxed{}}{\boxed{}} = \dfrac{\boxed{}}{\boxed{}}$

3 Solve the fraction calculations. Write answers greater than 1 as an improper fraction and a mixed number. Show any working out.

a $\dfrac{8}{9} + \dfrac{11}{13} = \dfrac{\boxed{}}{\boxed{}} + \dfrac{\boxed{}}{\boxed{}} = \dfrac{\boxed{}}{\boxed{}}$ or $\boxed{}$

b $\dfrac{17}{12} - \dfrac{10}{11} = \dfrac{\boxed{}}{\boxed{}} - \dfrac{\boxed{}}{\boxed{}} = \dfrac{\boxed{}}{\boxed{}}$

c $\dfrac{13}{12} + \dfrac{22}{15} = \dfrac{\boxed{}}{\boxed{}} + \dfrac{\boxed{}}{\boxed{}} = \dfrac{\boxed{}}{\boxed{}}$ or $\boxed{}$

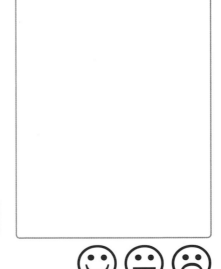

Date: _____

🙂 😐 🙁

Number

Lesson 3: **Multiplying fractions by whole numbers**

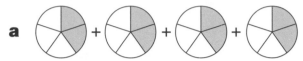

- Multiply proper fractions by whole numbers

1 Use the fraction models to find the sum.

a
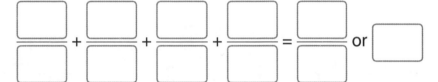

$$\frac{\boxed{}}{\boxed{}} + \frac{\boxed{}}{\boxed{}} + \frac{\boxed{}}{\boxed{}} + \frac{\boxed{}}{\boxed{}} = \frac{\boxed{}}{\boxed{}} \text{ or } \boxed{}$$

b
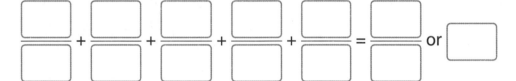

$$\frac{\boxed{}}{\boxed{}} + \frac{\boxed{}}{\boxed{}} + \frac{\boxed{}}{\boxed{}} + \frac{\boxed{}}{\boxed{}} + \frac{\boxed{}}{\boxed{}} = \frac{\boxed{}}{\boxed{}} \text{ or } \boxed{}$$

c

$$\frac{\boxed{}}{\boxed{}} + \frac{\boxed{}}{\boxed{}} + \frac{\boxed{}}{\boxed{}} = \frac{\boxed{}}{\boxed{}} \text{ or } \boxed{}$$

2 Use the models to multiply.

a $\frac{3}{10} \times 3 = \dfrac{\boxed{}}{\boxed{}}$

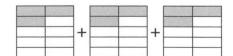

b $\frac{5}{8} \times 4 = \dfrac{\boxed{}}{\boxed{}} \text{ or } \boxed{}$

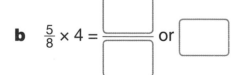

Number

c $\frac{3}{4} \times 6 = \dfrac{\boxed{}}{\boxed{}}$ or $\boxed{}$

d $\frac{4}{5} \times 5 = \dfrac{\boxed{}}{\boxed{}}$ or $\boxed{}$

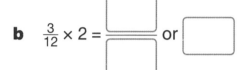

 3 Draw your own area models to multiply the fractions.

4

a $\frac{5}{6} \times 5 = \dfrac{\boxed{}}{\boxed{}}$ or $\boxed{}$

b $\frac{3}{12} \times 2 = \dfrac{\boxed{}}{\boxed{}}$ or $\boxed{}$

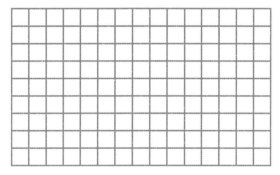

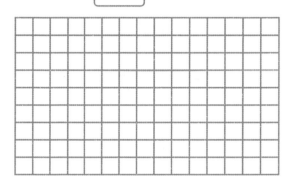

4 Solve the calculations.

a $\frac{5}{7} \times 3 = \dfrac{\boxed{}}{\boxed{}}$ or $\boxed{}$

b $\frac{7}{8} \times 6 = \dfrac{\boxed{}}{\boxed{}}$ or $\boxed{}$

c $\frac{2}{9} \times 5 = \dfrac{\boxed{}}{\boxed{}}$ or $\boxed{}$

d $\frac{5}{12} \times 2 = \dfrac{\boxed{}}{\boxed{}}$ or $\boxed{}$

e $\frac{9}{10} \times 6 = \dfrac{\boxed{}}{\boxed{}}$ or $\boxed{}$

f $\frac{7}{10} \times 5 = \dfrac{\boxed{}}{\boxed{}}$ or $\boxed{}$

g $\frac{13}{13} \times 8 = \dfrac{\boxed{}}{\boxed{}}$ or $\boxed{}$

h $\frac{17}{17} \times 6 = \dfrac{\boxed{}}{\boxed{}}$ or $\boxed{}$

Date: _____

☺ 😐 ☹

Number

Lesson 4: **Dividing fractions by whole numbers**

• Divide proper fractions by whole numbers

1 Use the grid to draw a diagram that models each fraction.

a $\frac{2}{3}$

b $\frac{7}{8}$

c $\frac{2}{5}$

d $\frac{3}{10}$

2 Use the models to divide.

a $\frac{2}{5} \div 4 = \dfrac{}{}$

b $\frac{3}{8} \div 5 = \dfrac{}{}$

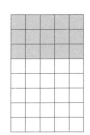

c $\frac{5}{6} \div 6 = \dfrac{}{}$

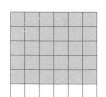

d $\frac{3}{7} \div 5 = \dfrac{}{}$

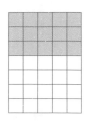

Number

 3 Draw your own area models to divide the fractions.

 4

a $\frac{3}{4} \div 6 =$ 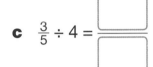 or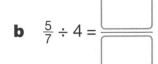

b $\frac{5}{7} \div 4 =$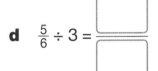

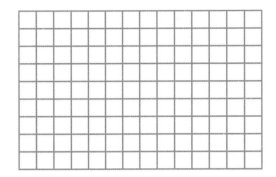

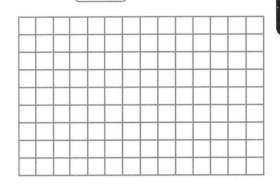

c $\frac{3}{5} \div 4 =$ ☐

d $\frac{5}{6} \div 3 =$ ☐

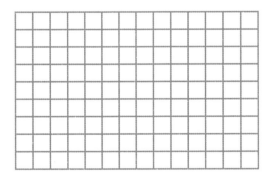

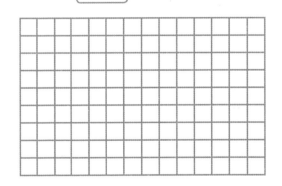

4 Solve the calculations.

a $\frac{2}{3} \div 3 =$ ☐

b $\frac{3}{8} \div 4 =$ ☐

c $\frac{5}{7} \div 3 =$ ☐

d $\frac{7}{8} \div 5 =$ ☐

e $\frac{3}{4} \div 8 =$ ☐

f $\frac{7}{9} \div 10 =$ ☐

Date: _____

Lesson 1: **Percentages of shapes**

- Recognise percentages of shapes

1 The shaded part of each 100 grid represents a percentage.
Write the percentage shown.

a
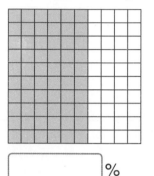

_____ %

b
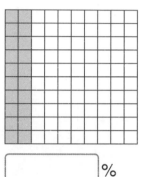

_____ %

c
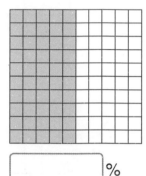

_____ %

d
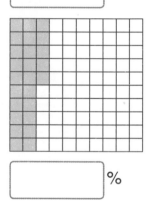

_____ %

e
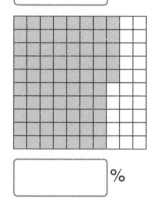

_____ %

f
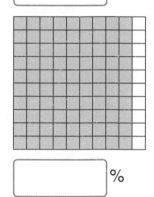

_____ %

2 Shade the squares in the grid to show each percentage.

a 5%

b 35%

c 85%

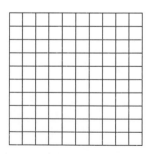
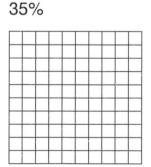
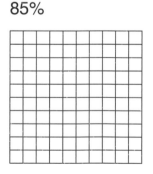

3 Write the percentage of each grid in **2** that is NOT shaded.

a _____ %   **b** _____ %   **c** _____ %

Number

4 Write the percentage, fraction and decimal shown.

a

[ ] % = [ ]/[ ] = [ ]

b

[ ] % = [ ]/[ ] = [ ]

c
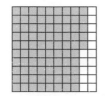

[ ] % = [ ]/[ ] = [ ]

d

[ ] % = [ ]/[ ] = [ ]

e
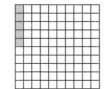

[ ] % = [ ]/[ ] = [ ]

f
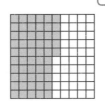

[ ] % = [ ]/[ ] = [ ]

5 Shade the squares of the grid to show each percentage.

a 45%   **b** 15%   **c** 85%

6 Write the fraction shaded of each grid in **5** in its simplest form.

a [ ]/[ ]   **b** [ ]/[ ]   **c** [ ]/[ ]

Date: _____

Lesson 2: **Percentages of whole numbers (1)**

Number

• Calculate percentages of whole numbers and quantities

1 Work out these percentages.

a 50% of 400 ⬚   **b** 50% of $440 ⬚

c 50% of 680 kg ⬚   **d** 50% of 128 km ⬚

e 50% of $890 ⬚   **f** 50% of 780 g ⬚

g 25% of 164 ml ⬚   **h** 25% of $3200 ⬚

i 25% of 840 m ⬚   **j** 25% of $920 ⬚

k 25% of 460 g ⬚   **l** 25% of 76 cm ⬚

m 10% of 490 g ⬚   **n** 10% of 3430 l ⬚

2 Complete the table. Convert percentages to fractions to work out the amounts.

	5% of	25% of	75% of
$20			
$120			
$4			

3 Work out the percentages of the amount.

If 100% is 2800, what is:
1% ⬚   10% ⬚   5% ⬚   50%? ⬚

Number

4 Use the answers in **3** to find the following percentages. Show your working in each box.

a What is 15% of 2800?

b What is 55% of 2800?

c What is 25% of 2800?

d What is 45% of 2800?

e What is 75% of 2800?

f What is 95% of 2800?

5 Solve these percentage problems. Show your working out.

a There are 90 children in a playground. 30% are girls. How many girls are there?

b There are 130 counters in a bag. 70% are red. How many counters are red?

c Kam has collected 124 football stickers for his book. He gives 25% of them to his friend. How many does he give away?

d Melissa decides to donate 15% of her books to a charity shop. She has 120 books. How many books does she donate?

Date: _____

Number

Lesson 3: **Percentages of whole numbers (2)**

- Calculate percentages of whole numbers and quantities

1 Complete the table by calculating the percentages of the amounts.

	10% of	20% of	25% of	50% of	75% of
$10					
$20					
$60					
$140					

2 Calculate the new prices.

	Rise: 10%	Discount: 20%	Rise: 25%	Discount: 50%	Rise: 75%
$40					
$200					
$800					
$2400					

3 Learners travel to school in different ways.

If 60 learners travel to school, how many will travel by:

Type of travel	Percentage
Walk	45%
Bus	20%
Car	25%
Bicycle	10%

a walking? ☐   **b** bus? ☐

c car? ☐   **d** bicycle? ☐

Show your working in the box provided.

4 Solve these percentage problems.

a A mobile phone is advertised as '5% off'. The original price of the phone was $400. What is the sale price?

b Tom's bill for a meal in his local café is $42. He gives the waiter a tip of 15%. How much is the tip?

c Mary bought a painting for $640. A year later she sold it, making 35% profit. How much did she sell the painting for?

d A car is advertised as '40% off'. The original price of the car was $17 600. What is the sale price?

Date: _____

Number

Lesson 4: **Comparing percentages**

• Compare and order percentages of quantities

1 Write the percentages in order, from least to greatest. Assume each set of percentages refers to the same quantity.

a 37%   56%   19%   48%   _____

b 45%   54%   44%   51%   _____

c 76%   67%   74%   68%   _____

d 85%   95%   82%   59%   _____

e 72%   65%   27%   56%   57%   _____

f 34%   43%   33%   42%   32%   _____

 Draw a ring around the greater number.

a   0.4 or 50%   **b**   60% or $\frac{1}{2}$ **c**   0·8 or $\frac{7}{10}$

d $\frac{1}{4}$ or 30%   **e**   70% or 0·6   **f**   0·2 or $\frac{1}{10}$

g   20% or 0·1   **h** $\frac{3}{4}$ or 0·7   **i** $\frac{3}{10}$ or 25%

 Write each set of numbers in order, from smallest to greatest.

a 60%, $\frac{1}{2}$, 0·7   Order: [   ] [   ] [   ]

b $\frac{7}{10}$, 0·6, 50%   Order: [   ] [   ] [   ]

c 0·3, 20%, $\frac{1}{10}$ Order: [   ] [   ] [   ]

d 90%, $\frac{5}{5}$, 0·8   Order: [   ] [   ] [   ]

e $\frac{1}{4}$, 20%, 0·3   Order: [   ] [   ] [   ]

f 0·9, $\frac{3}{4}$, 70%   Order: [   ] [   ] [   ]

Number

4 Order the amounts from the smallest amount to the largest amount. Write the letter codes in order.

a   **A:** $\frac{1}{2}$ of a bag of sugar   **B:** 40% of a bag of sugar   **C:** 0·3 of a bag of sugar

Order: ☐ < ☐ < ☐

b   **A:** 90% of a glass of juice   **B:** 0·8 of a glass of juice   **C:** $\frac{3}{4}$ of a glass of juice

Order: ☐ < ☐ < ☐

c   **A:** $\frac{2}{5}$ of a cake   **B:** 20% of a cake   **C:** 0·3 of a cake

Order: ☐ < ☐ < ☐

c   **A:** 0·9 of a chocolate bar   **B:** $\frac{4}{5}$ of a chocolate bar   **C:** 70% of a chocolate bar

Order: ☐ < ☐ < ☐

5 Fill in the missing numbers to make the statements true. Each statement must include at least a decimal, percentage and a fraction.

a   20% < ☐ < $\frac{2}{5}$ **b** $\frac{7}{10}$ > ☐ > 0·5

c   0·8 < ☐ < 100 %   **d** $\frac{3}{5}$ < ☐ < 80%

e   0·1 > ☐ > $\frac{3}{10}$ **f**   0·3 < ☐ < 50%

g   60% < ☐ < $\frac{4}{5}$ **h**   0·9 > ☐ > 70%

i   70% < ☐ < 0·8

Date: _____

Lesson 1: **Adding decimals (mental strategies)**

- Add pairs of decimals mentally

1 Add the decimals mentally.

a 0·6 + 0·6 = [　　]

b 0·8 + 0·3 = [　　]

c 0·5 + 0·8 = [　　]

d 0·3 + 0·9 = [　　]

e 0·7 + 0·8 = [　　]

f 0·6 + 0·9 = [　　]

g 0·8 + 0·12 = [　　]

h 0·7 + 0·25 = [　　]

i 1·5 + 0·52 = [　　]

j 1·6 + 0·54 = [　　]

k 1·7 + 0·61 = [　　]

l 1·7 + 0·4 = [　　]

2 Add the numbers mentally using any strategy you prefer. Show your working in the box provided.

a 2·6 + 0·45 = [　　]

b 3·4 + 0·83 = [　　]

c 0·37 + 0·99 = [　　]

d 0·53 + 0·78 = [　　]

Number

e $4 \cdot 3 + 0 \cdot 86 =$ ▢

f $6 \cdot 6 + 0 \cdot 74 =$ ▢

3 Lucas says:

I find the sum of $0 \cdot 76 + 0 \cdot 58$ by adding $0 \cdot 6$ and then adding $0 \cdot 02$.

Is Lucas correct? ▢

If not, what mistake has he made and how would you correct his thinking?

_____

_____

4 Answer the word problems.

a Ling pours $5 \cdot 4$ litres of orange juice into a large container. He then adds $0 \cdot 74$ litres of lemonade.
What is the total amount of liquid in the container? ▢ l

b Bhavna runs $7 \cdot 8$ kilometres. Then she walks a further $0 \cdot 56$ kilometres.
What distance has she travelled altogether? ▢ km

5 Fill in the missing numbers.

a $0 \cdot 55 +$ ▢ $= 0 \cdot 83$

b ▢ $+ 0 \cdot 72 = 2 \cdot 12$

c ▢ $+ 0 \cdot 86 = 3 \cdot 36$

d $0 \cdot 77 +$ ▢ $= 0 \cdot 96$

e $0 \cdot 67 +$ ▢ $= 1 \cdot 16$

f ▢ $+ 0 \cdot 87 = 5 \cdot 47$

g ▢ $+ 0 \cdot 53 = 8 \cdot 23$

h $0 \cdot 87 +$ ▢ $= 1 \cdot 36$

i $0 \cdot 95 +$ ▢ $= 1 \cdot 83$

j ▢ $+ 0 \cdot 77 = 9 \cdot 57$

Date: _____

Lesson 2: **Adding decimals (written methods)**

• Add pairs of decimals using written methods

1 Find two addends that make the sum. The first one has been done for you.

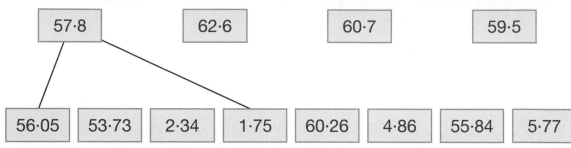

| 57·8 | | 62·6 | | 60·7 | | 59·5 |

| 56·05 | 53·73 | 2·34 | 1·75 | 60·26 | 4·86 | 55·84 | 5·77 |

2 Calculate using the expanded written method. Estimate the answer first.

a 25·48 + 3·765 = [   ]

Estimate: [   ]

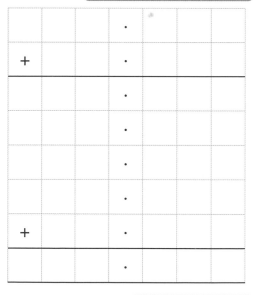

b 47·65 + 5·864 = [   ]

Estimate: [   ]

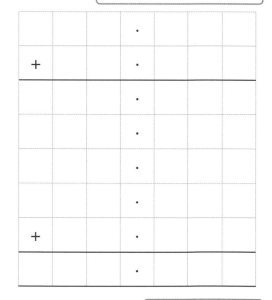

c 66·96 + 8·852 = [   ]

Estimate: [   ]

d 78·87 + 7·648 = [   ]

Estimate: [   ]

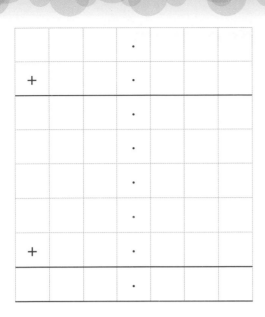

3 Solve using the formal written method. Estimate the answer first.

a $18.86 + 7.485 =$ [   ]   **b** $33.97 + 8.254 =$ [   ]

    Estimate: [   ]   Estimate: [   ]

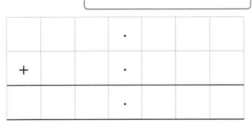 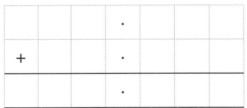

4   Use repeated addition to solve each problem. The first one is done for you.

a $38.57 + 2 \times 3.348 =$ **b** $25.38 + 2 \times 4.546 =$

> $38.57 + 3.348 + 3.348$
> $= 41.918 + 3.348$
> $= 45.266$

c $64.66 + 2 \times 8.747 =$ **d** $48.485 + 3 \times 3.486 =$

Date: _____

Number

Lesson 3: **Subtracting decimals (mental strategies)**

• Subtract pairs of decimals mentally

1 Subtract the decimals mentally.

a $2 \cdot 8 - 0 \cdot 6 =$ ⬜

b $7 \cdot 8 - 0 \cdot 3 =$ ⬜

c $5 \cdot 5 - 1 \cdot 1 =$ ⬜

d $9 \cdot 3 - 5 \cdot 2 =$ ⬜

e $8 \cdot 7 - 4 \cdot 4 =$ ⬜

f $6 \cdot 6 - 3 \cdot 2 =$ ⬜

g $9 \cdot 8 - 7 \cdot 5 =$ ⬜

h $5 \cdot 7 - 5 \cdot 5 =$ ⬜

i $8 \cdot 8 - 8 \cdot 1 =$ ⬜

j $6 \cdot 3 - 6 \cdot 2 =$ ⬜

k $2 \cdot 7 - 2 \cdot 5 =$ ⬜

l $3 \cdot 9 - 3 \cdot 7 =$ ⬜

2 Subtract the numbers mentally, using any strategy you prefer. Show your working in the box provided.

a $4 \cdot 6 - 1 \cdot 45 =$ ⬜

b $8 \cdot 37 - 0 \cdot 9 =$ ⬜

c $6 \cdot 373 - 6 \cdot 35 =$ ⬜

d $5 \cdot 6 - 4 \cdot 23 =$ ⬜

114

Number

e $7 \cdot 8 - 3 \cdot 49 =$ ⬚

f $9 \cdot 884 - 9 \cdot 87 =$ ⬚

3 Model the subtraction of $7 \cdot 4 - 3 \cdot 86$ using place value counters.

4 Use a convincing argument to prove that the answer is $3 \cdot 54$.

Step 1

1s	• 0.1s	0.01s
① ① ① ① ① ① ①	0·1 0·1 0·1 0·1	

Step 2

1s	• 0.1s	0.01s

Step 3

1s	• 0.1s	0.01s

4 Fill in the missing numbers.

a $5 \cdot 5 -$ ⬚ $= 4 \cdot 25$

b   ⬚ $- 4 \cdot 36 = 3 \cdot 44$

c   ⬚ $- 0 \cdot 9 = 5 \cdot 38$

d $8 \cdot 4 -$ ⬚ $= 6 \cdot 21$

e $4 \cdot 453 -$ ⬚ $= 0 \cdot 043$

f   ⬚ $- 9 \cdot 63 = 0 \cdot 048$

g   ⬚ $- 3 \cdot 35 = 4 \cdot 15$

h $3 \cdot 6 -$ ⬚ $= 2 \cdot 36$

i $9 \cdot 57 -$ ⬚ $= 8 \cdot 67$

j   ⬚ $- 3 \cdot 29 = 2 \cdot 41$

k   ⬚ $- 6 \cdot 52 = 0 \cdot 066$

l $1 \cdot 645 -$ ⬚ $= 0 \cdot 035$

Date: _____

Number

Lesson 4: **Subtracting decimals (written methods)**

• Subtract pairs of decimals using written methods

1 Work out these subtraction calculations.

a	3·8	**b**	44·7	**c**	4·85	**d**	7·9
	− 2·6		− 22·5		− 2·62		− 5·4

e	68·8	**f**	7·77	**g**	9·9	**h**	99·9
	− 42·5		− 2·22		− 5·8		− 15·6

2 Solve using the formal written method. Both minuend and subtrahend have the same number of digits. Estimate the answer first.

a 56·3 − 42·7 = [   ]

Estimate: [   ]

b 9·73 − 4·37 = [   ]

Estimate: [   ]

c 96·54 − 48·26 = [   ]

Estimate: [   ]

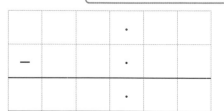

d 75·44 − 58·72 = [   ]

Estimate: [   ]

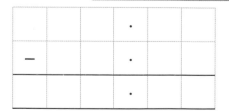

Number

3 Solve using the formal written method. The minuend and subtrahend have a different number of digits. Estimate the answer first.

a   9·3 − 4·26 = [ ]

Estimate: [ ]

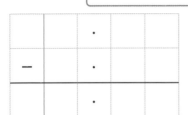

b   66·445 − 8·82 = [ ]

Estimate: [ ]

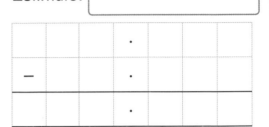

c   8·5 − 6·28 = [ ]

Estimate: [ ]

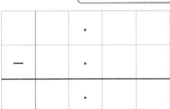

d   92·376 − 7·55 = [ ]

Estimate: [ ]

4 Solve using the formal written method. The subtractions involve the regrouping of more columns than previous questions.

a   8·746 − 2·858 = [ ]

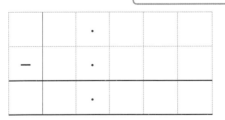

b   43·23 − 18·67 = [ ]

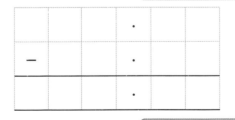

c   7·346 − 4·968 = [ ]

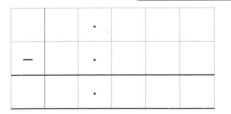

d   94·67 − 77·88 = [ ]

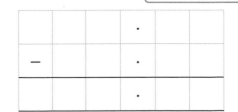

Date: _____

Lesson 1: **Multiplying decimals by 1-digit whole numbers (1)**

• Multiply decimals by 1-digit whole numbers

You will need
• paper for working out

1 Decompose the numbers into tens, ones, tenths and hundredths.

Example:

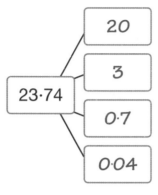

23·74 → 20, 3, 0·7, 0·04

a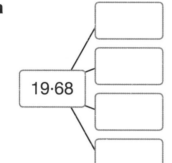

19·68

b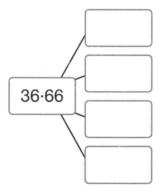

36·66

c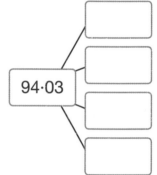

94·03

d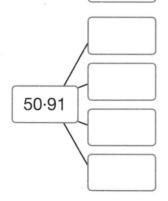

50·91

e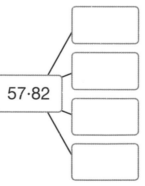

57·82

f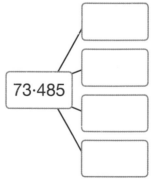

73·485

 Draw the arrangement of place value counters to show each calculation and then use it to find the product. Estimate the answer first.

a 31·2 × 7 = [   ]

b 2·43 × 5 = [   ]

Estimate: [   ]

Estimate: [   ]

Number

	10s	1s	0.1s
			•
Total			•

	1s	0.1s	0.01s
		•	
Total		•	

3 Use the grid method to multiply. Estimate the answer first.

a 27·2 × 6 = [   ]

Estimate: [   ]

× [   ] [   ] [   ]

[   ] [   ] [   ] [   ]

b 198·4 × 3 = [   ]

Estimate: [   ]

× [   ] [   ] [   ] [   ]

[   ] [   ] [   ] [   ]

c 7·26 × 7 = [   ]

Estimate: [   ]

× [   ] [   ] [   ]

[   ] [   ] [   ] [   ]

d 83·48 × 7 = [   ]

Estimate: [   ]

× [   ] [   ] [   ]

[   ] [   ] [   ] [   ]

4 Draw a line to match each calculation to its product. Show your working out on a separate piece of paper.

97·44 × 7	86·3 × 8	98·57 × 6	137·5 × 5

687·5	682·08	690·4	591·42

Date: _____

119

Number

Lesson 2: **Multiplying decimals by 1-digit whole numbers (2)**

• Multiply decimals by 1-digit whole numbers

1 Use partitioning to multiply. Estimate the answer first.

a 28·7 × 6 = [ ]

Estimate: [ ]

b 8·48 × 7 = [ ]

Estimate: [ ]

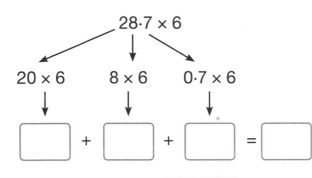

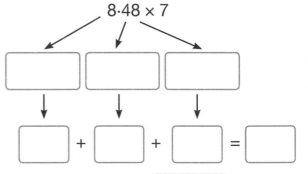

c 9·3 × 80 = [ ]

Estimate: [ ]

d 15·21 × 50 = [ ]

Estimate: [ ]

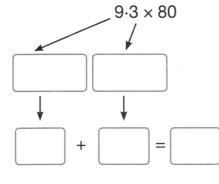

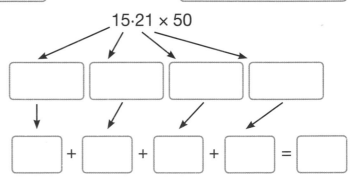

2 Use the expanded written method to multiply. Estimate the answer first.

a 63·7 × 6 = [ ] × 6 ÷ 10

Estimate: [ ]

b 75·36 × 7 = [ ] × 7 ÷ 100

Estimate: [ ]

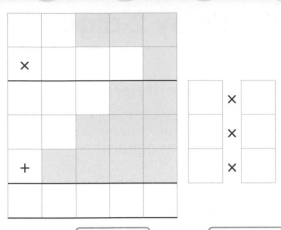

 ×

Answer: [ ] ÷ 10 = [ ]

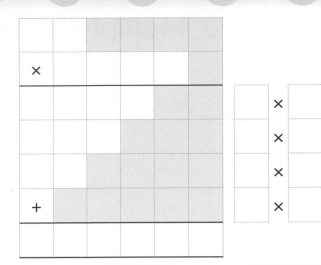

Answer: [ ] ÷ 100 = [ ]

3 Use the formal written method to multiply. Use the squared grid to plan your working out. Estimate the answer first.

a 5·8 × 8 = [ ] × 8 ÷ 10

Estimate: [ ]

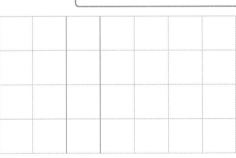

Answer: [ ] ÷ 10 = [ ]

b 565·7 × 4 = [ ] × 4 ÷ 10

Estimate: [ ]

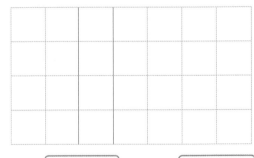

Answer: [ ] ÷ 10 = [ ]

4 Use any mental or written method to solve these calculations.

a 878·8 × 80 = [ ]

b 84·76 × 90 = [ ]

Date: _____

121

Number

Lesson 3: **Multiplying decimals by 2-digit whole numbers (1)**

• Multiply decimals by 2-digit whole numbers

You will need

• paper for working out

1 Complete the multiplications. Look carefully to spot the pattern.

a $5 \times 8 =$ ☐

b $5 \times 80 =$ ☐

c $5 \times 800 =$ ☐

d $5 \times 0.8 =$ ☐

e $5 \times 0.08 =$ ☐

f $50 \times 0.08 =$ ☐

g $4 \times 7 =$ ☐

h $4 \times 70 =$ ☐

i $4 \times 700 =$ ☐

j $4 \times 0.7 =$ ☐

k $4 \times 0.07 =$ ☐

l $40 \times 0.07 =$ ☐

m $8 \times 6 =$ ☐

n $8 \times 60 =$ ☐

o $8 \times 600 =$ ☐

p $8 \times 0.6 =$ ☐

q $8 \times 0.06 =$ ☐

r $80 \times 0.06 =$ ☐

s $6 \times 9 =$ ☐

t $6 \times 90 =$ ☐

u $6 \times 900 =$ ☐

v $6 \times 0.9 =$ ☐

w $6 \times 0.09 =$ ☐

x $60 \times 0.09 =$ ☐

2 Use the grid method to multiply. Estimate the answer first.

a $0.8 \times 34 =$ ☐

Estimate: ☐

b $6.6 \times 26 =$ ☐

Estimate: ☐

c   53·4 × 47 = [ ]

Estimate: [ ]

   + _____

_____

d   474·5 × 38 = [ ]

Estimate: [ ]

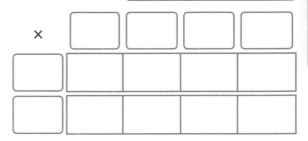

   + _____

_____

e   7·34 × 67 = [ ]

Estimate: [ ]

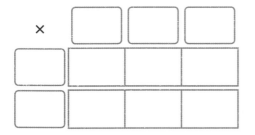

   + _____

_____

f   33·46 × 42 = [ ]

Estimate: [ ]

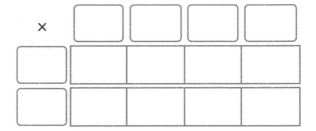

   + _____

_____

3   Draw a line to match each calculation to its product. Show your working out on a separate piece of paper.

| 8·4 × 38 | 6·37 × 56 | 47·6 × 24 | 73·44 × 16 |

| 1142·4 | 1175·04 | 319·2 | 356·72 |

Date: _____

Number

Lesson 4: **Multiplying decimals by 2-digit whole numbers (2)**

• Multiply decimals by 2-digit whole numbers

1 Use partitioning to multiply. Estimate the answer first.

a 0·8 × 34 = ☐

Estimate: ☐

0·8 × 34 = (0·8 × ☐) +

(0·8 × ☐)

= ☐ + ☐

= ☐

b 6·7 × 17 = ☐

Estimate: ☐

6·7 × 17 = (6 × ☐) +

(0·7 × ☐)

= ☐ + ☐

= ☐

2 Use the expanded written method to multiply. Estimate the answer first.

a 84·5 × 63 = ☐ × 63 ÷ 10

Estimate: ☐

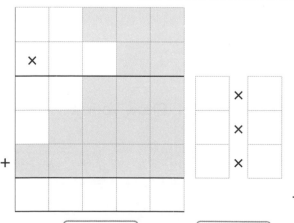

Answer: ☐ ÷ 10 = ☐

b 24·74 × 43 = ☐ × 43 ÷ 100

Estimate: ☐

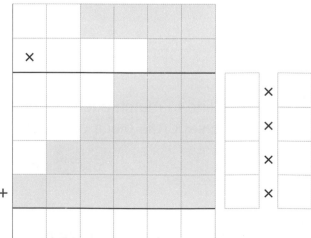

Answer: ☐ ÷ 100 = ☐

Number

3 Use the formal written method to multiply. Estimate the answer first.

a 263·7 × 4 = [ ] × 4 ÷ 10   **b** 5·86 × 7 = [ ] × 7 ÷ 100

Estimate: [ ]   Estimate: [ ]

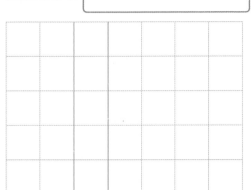 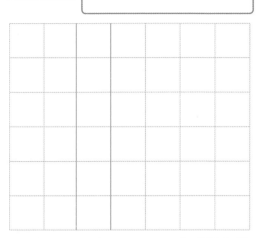

Answer: [ ] ÷ 10 = [ ]   Answer: [ ] ÷ 100 = [ ]

4 Use any mental or written method to solve these calculations.

a 868·8 × 87 = [ ]   **b** 93·67 × 93 = [ ]

c 9·66 × 97 = [ ]   **d** 947·7 × 94 = [ ]

5 Compare the method(s) you used to solve the calculations in **4** with a
7 classmate. Which method is the most efficient? Why?

_____

_____

Date: _____

Lesson 1: **Dividing one-place decimals by whole numbers (1)**

- Divide one-place decimals by whole numbers

You will need
- paper for working out

1 Use known division facts to solve these problems.

a 24 ÷ 2 = ☐

 2·4 ÷ 2 = ☐

b 33 ÷ 3 = ☐

 3·3 ÷ 3 = ☐

c 48 ÷ 4 = ☐

 4·8 ÷ 4 = ☐

d 46 ÷ 2 = ☐

 4·6 ÷ 2 = ☐

e 66 ÷ 3 = ☐

 6·6 ÷ 3 = ☐

f 55 ÷ 5 = ☐

 5·5 ÷ 5 = ☐

2 Estimate (E) first and then use the expanded written method of division to work out the answer to each calculation.

a 6·8 ÷ 4 = ☐ E: ☐

6·8 ÷ 4 is equivalent to

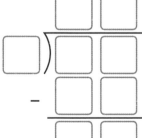

☐ ÷ 10 = ☐

b 8·4 ÷ 6 = ☐ E: ☐

8·4 ÷ 6 is equivalent to

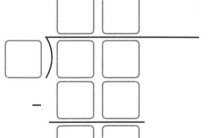

☐ ÷ 10 = ☐

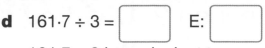

Number

c 71·2 ÷ 4 = [ ] E: [ ]

71·2 ÷ 4 is equivalent to

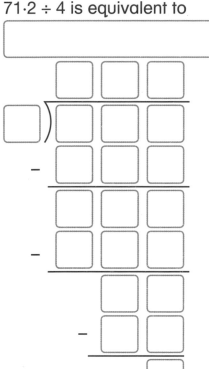

[ ] ÷ 10 = [ ]

d 161·7 ÷ 3 = [ ] E: [ ]

161·7 ÷ 3 is equivalent to

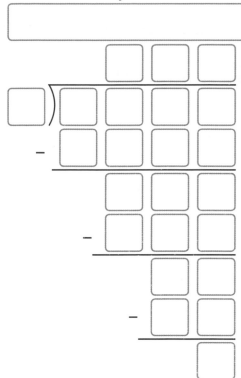

[ ] ÷ 10 = [ ]

3 Estimate first and then use the expanded written method of division to work out the answer to each calculation. Complete your working out on a separate sheet of paper.

a 7·5 ÷ 3 = [ ]

E: [ ]

b 6·2 ÷ 4 = [ ]

E: [ ]

c 9·6 ÷ 6 = [ ]

E: [ ]

d 75·6 ÷ 4 = [ ]

E: [ ]

e 87·6 ÷ 6 = [ ]

E: [ ]

f 94·2 ÷ 3 = [ ]

E: [ ]

g 194·4 ÷ 3 = [ ]

E: [ ]

h 241·5 ÷ 5 = [ ]

E: [ ]

i 395·2 ÷ 4 = [ ]

E: [ ]

Date: _____

127

Lesson 2: **Dividing one-place decimals by whole numbers (2)**

Number

• Divide one-place decimals by whole numbers

1 Partition and solve. The first one is done for you.

Example:

16·4 ÷ 4 = | 16 ÷ 4 | + | 0·4 ÷ 4 |

= | 4 | + | 0·1 |

= | 4·1 |

a 20·5 ÷ 5 = [ ] + [ ]

= [ ] + [ ]

= [ ]

b 15·6 ÷ 3 = [ ] + [ ]

= [ ] + [ ]

= [ ]

c 30·6 ÷ 6 = [ ] + [ ]

= [ ] + [ ]

= [ ]

2 Estimate (E) first and then use the short division method to work out the answer to each calculation.

a 9·6 ÷ 8 = [ ]   E: [ ]

9·6 ÷ 8 is equivalent to

| 96 ÷ 8 ÷ 10 |

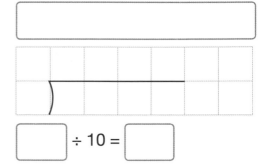

[ ] ÷ 10 = [ ]

b 9·8 ÷ 5 = [ ]   E: [ ]

9·8 ÷ 5 is equivalent to

[ ]

[ ] ÷ 10 = [ ]

c 73·6 ÷ 4 = [ ]   E: [ ]

73·6 ÷ 4 is equivalent to

[ ]

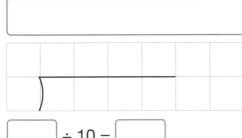

[ ] ÷ 10 = [ ]

d 77·4 ÷ 6 = [ ]   E: [ ]

77·4 ÷ 6 is equivalent to

[ ]

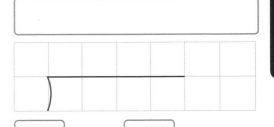

[ ] ÷ 10 = [ ]

e 253·2 ÷ 4 = [ ]   E: [ ]

253·2 ÷ 4 is equivalent to

[ ]

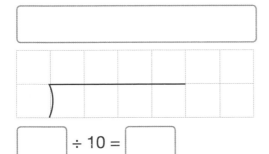

[ ] ÷ 10 = [ ]

f 476·7 ÷ 6 = [ ]   E: [ ]

476·7 ÷ 6 is equivalent to

[ ]

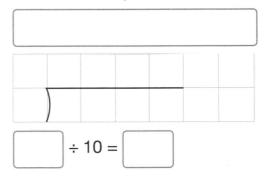

[ ] ÷ 10 = [ ]

3 Estimate (E) first and then use the short division method to work out the answer to each calculation.

a 554.4 ÷ 7 = [ ]   E: [ ]

554.4 ÷ 7 is equivalent to

[ ]

[ ] ÷ 10 = [ ]

b 796.4 ÷ 8 = [ ]   E: [ ]

796.4 ÷ 8 is equivalent to

[ ]

[ ] ÷ 10 = [ ]

Date: _____

Number

Lesson 3: **Dividing two-place decimals by whole numbers (1)**

• Divide two-place decimals by whole numbers

1 Use known division facts to solve these problems.

a $846 \div 2 = \boxed{}$ **b** $636 \div 3 = \boxed{}$ **c** $848 \div 4 = \boxed{}$

$8{\cdot}46 \div 2 = \boxed{}$ $6{\cdot}36 \div 3 = \boxed{}$ $8{\cdot}48 \div 4 = \boxed{}$

d $505 \div 5 = \boxed{}$ **e** $606 \div 6 = \boxed{}$ **f** $969 \div 3 = \boxed{}$

$5{\cdot}05 \div 5 = \boxed{}$ $6{\cdot}06 \div 6 = \boxed{}$ $9{\cdot}69 \div 3 = \boxed{}$

2 Estimate (E) first and then use the expanded written method of division to work out the answer to each calculation.

a $18{\cdot}27 \div 3 = \boxed{}$ E: $\boxed{}$ **b** $56{\cdot}16 \div 8 = \boxed{}$ E: $\boxed{}$

$18{\cdot}27 \div 3$ is equivalent to $56{\cdot}16 \div 8$ is equivalent to

$\boxed{} \div 100 = \boxed{}$ $\boxed{} \div 100 = \boxed{}$

c $48.24 \div 12 =$ [ ] E: [ ]

48·24 ÷ 12 is equivalent to

[ ]

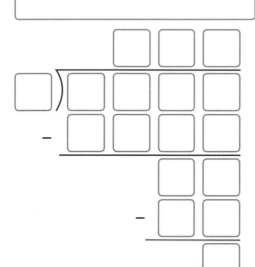

[ ] ÷ 100 = [ ]

d $75.45 \div 15 =$ [ ] E: [ ]

75·45 ÷ 15 is equivalent to

[ ]

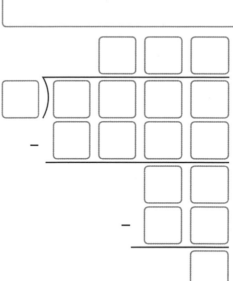

[ ] ÷ 100 = [ ]

3 Estimate first and then use the expanded written method of division to work out the answer to each calculation. Show your working out in the boxes.

a $78.72 \div 6 =$ [ ]

Estimate: [ ]

b $84.91 \div 7 =$ [ ]

Estimate: [ ]

c $96.8 \div 8 =$ [ ]

Estimate: [ ]

Date: _____

Number

Lesson 4: **Dividing two-place decimals by whole numbers (2)**

• Divide two-place decimals by whole numbers

1 Partition and solve. The first one is done for you.

Example:

$20.84 \div 4 = \boxed{20 \div 4} + \boxed{0.84 \div 4}$

$= \boxed{5} + \boxed{0.21}$

$= \boxed{5.21}$

a $15.25 \div 5 = \boxed{} + \boxed{}$

$= \boxed{} + \boxed{}$

$= \boxed{}$

b $9.27 \div 3 = \boxed{} + \boxed{}$

$= \boxed{} + \boxed{}$

$= \boxed{}$

c $18.42 \div 6 = \boxed{} + \boxed{}$

$= \boxed{} + \boxed{}$

$= \boxed{}$

2 Estimate (E) first and then use the short division method to work out the answer to each calculation.

a $9.66 \div 3 = \boxed{}$ E: $\boxed{}$

$9.66 \div 3$ is equivalent to

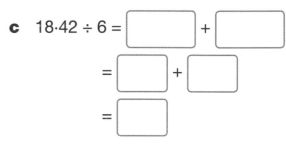

$966 \div 3 \div 100$

$\boxed{} \div 100 = \boxed{}$

b $9.85 \div 5 = \boxed{}$ E: $\boxed{}$

$9.85 \div 5$ is equivalent to

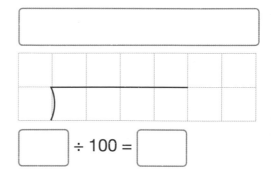

$\boxed{} \div 100 = \boxed{}$

Number

c   24·36 ÷ 4 = ☐   E: ☐

24·36 ÷ 4 is equivalent to

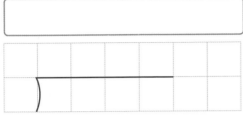

☐ ÷ 100 = ☐

d   28·42 ÷ 7 = ☐   E: ☐

28·42 ÷ 7 is equivalent to

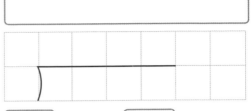

☐ ÷ 100 = ☐

e   48·72 ÷ 8 = ☐   E: ☐

48·72 ÷ 8 is equivalent to

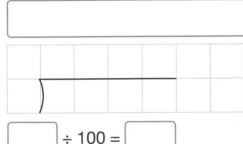

☐ ÷ 100 = ☐

f   54·12 ÷ 6 = ☐   E: ☐

54·12 ÷ 6 is equivalent to

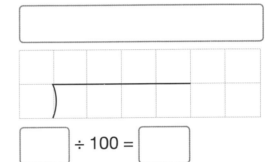

☐ ÷ 100 = ☐

3   Estimate (E) first and then use the short division method to work out the answer to each calculation.

a   84·91 ÷ 7 = ☐   E: ☐

84·91 ÷ 7 is equivalent to

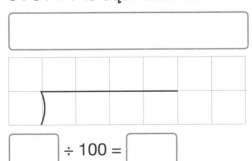

☐ ÷ 100 = ☐

b   92·64 ÷ 4 = ☐   E: ☐

92·64 ÷ 4 is equivalent to

☐ ÷ 100 = ☐

Date: _____

Number

Lesson 1: **Direct proportion (1)**

- Understand the relationship between two quantities when they are in direct proportion

1 Class 6 are playing basketball. Here are the number of shots each player has had and the number of baskets they scored.

Name	Number of shots in total	Number of baskets scored
Krishna	4	1
Alex	100	10
Lucy	6	3
Megan	3	1
Brett	16	8
Satpal	9	8

a Write the proportion of baskets scored for each player. Remember that proportion is a way to compare a part (the number of baskets) with the whole (the total number of shots).

Krishna scores $\dfrac{1}{4}$ of her shots.   Alex scores $\dfrac{}{}$ of his shots.

Lucy scores $\dfrac{}{}$ of her shots.   Megan scores $\dfrac{}{}$ of her shots.

Brett scores $\dfrac{}{}$ of his shots.   Satpal scores $\dfrac{}{}$ of his shots.

b Two players score half of the shots they take. Which players are they?

[   ] and [   ]

c Look at the proportions you wrote in Question **1** **a**. Which player do you think is the best at basketball? [   ] Why?

_____

_____

Number

2 Direct proportion means that, as one thing increases, so does another.

60c

50c for 3

20c for a pack of 10

a How much do two notepads cost? ☐

b How much will it cost to buy 9 pencils? ☐ Why?

_____

_____

c Tom spends 80c on sticky notes. How many packs does he buy? ☐ Why?

_____

_____

3 The trees are in proportion.

a How many times larger is the tall tree compared to the short tree? ☐

b What is the height of the trunk of the short tree? ☐ m

8 m
?m

32 m
18 m

4 These two triangles are similar.

a How many times larger is the triangle on the right compared to the triangle on the left? ☐

b Work out the length of side:

x: ☐ cm   y: ☐ cm

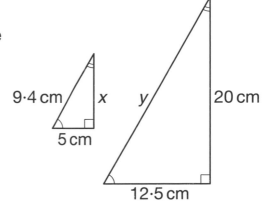
9·4 cm   x   y   20 cm
5 cm
12·5 cm

Date: _____

Lesson 2: **Direct proportion (2)**

Number

- Understand the relationship between two quantities when they are in direct proportion

You will need
- ruler

1 Katie wants to draw a picture of some animals to the right proportions, so she finds their heights.

Complete each sentence to help Katie compare the animals using proportion.

Name of animal	Height in cm (approx.)
Mouse	10 cm
Meerkat	30 cm
Lion	100 cm
Elephant	300 cm
Giraffe	500 cm

Example: About ⬚ 10 ⬚ mice are the same height as a lion. So a mouse is

$\dfrac{1}{10}$ of the height of a lion.

a About ⬚ mice are the same height as a meerkat.

So a mouse is $\dfrac{⬚}{⬚}$ of the height of a meerkat.

b About ⬚ lions are the same height as a giraffe. So a lion is $\dfrac{⬚}{⬚}$ of the height of a giraffe.

c About ⬚ lions are the same height as an elephant. So a lion is

$\dfrac{⬚}{⬚}$ of the height of an elephant.

d About ⬚ meerkats are the same height as an elephant.

So a meerkat is $\dfrac{⬚}{⬚}$ of the height of an elephant.

Number

 2 Answer the following word problems. Assume the relationship between the quantities is directly proportional.

a Louis buys 4 pens for $7. How many pens can he buy for:

i $14? ☐ pens **ii** $28? ☐ pens **iii** $3.50? ☐ pens

b Alisha buys 8 cards for $12. How many cards can she buy for:

i $36? ☐ cards **ii** $60? ☐ cards **iii** $3? ☐ cards

c Manish buys 8 notepads for $40. What is the price for:

i 24 notepads? **ii** 72 notepads? **iii** 1 notepad?

$☐ $☐ $☐

3 Use the recipe to work out the amounts.

a How much butter is required for 64 gingerbread men? ☐ g

b How much flour is required for 80 gingerbread men? ☐ g

c How many gingerbread men can be made with 15 g sugar (plus other ingredients)? ☐ gingerbread men

d How many gingerbread men can be made with 5 g ginger (plus other ingredients)? ☐ gingerbread men

> **Gingerbread Men**
>
> Makes 16 gingerbread
>
> 180 g flour   40 g ginger
>
> 110 g butter   30 g sugar

4 Ling asks a printing company to enlarge a photograph of her family to make a poster.

a Compared to the photograph, how many times larger did the printing company make the poster? ☐

b What is the length of the poster? ☐ cm

? cm

310·4 cm

37 cm

32 cm

Date: _____

☺ 😐 ☹

Lesson 3: **Equivalent ratios (1)**

• Use equivalent ratios to calculate unknown amounts

1 Class 6 take part in a Maths quiz. The table shows the number of points 6 learners score.

Compare the scores by writing ratios.

Then write the ratio in its simplest form.

Name	Points scored
Leon	4
Carla	10
Finn	2
Holly	16
Abdul	6
May	8

a The ratio of Finn's score to Leon's score.

☐ : ☐ = ☐ : ☐

b The ratio of Abdul's score to Finn's score.

☐ : ☐ = ☐ : ☐

c The ratio of May's score to Carla's score.

☐ : ☐ = ☐ : ☐

d The ratio of Leon's score to Holly's score.

☐ : ☐ = ☐ : ☐

2 Draw a ring around the ratios that are equivalent to the ratio in each box.

a | **2:3** | 4:8   6:9   10:15   9:6   14:23   24:36

b | **5:2** | 10:4   20:12   20:8   6:15   70:30   45:18

c | **3:8** | 18:53   21:49   15:40   60:160   23:72   36:96

d | **9:2** | 27:7   45:10   8:36   63:14   108:26   135: 30

Number

3 Answer each question, showing your working.

a The ratio of cows to sheep in a field is 2:3. If there are 10 cows, how many sheep are there?

b The ratio of adults to children on a bus is 7:4. If there are 20 children, how many adults are there?

c The ratio of herons to storks around a lake is 11:7. If there are 165 herons around the lake, how many storks are there?

4 Sidney prepares a fruit drink for a party.

For every 60 ml of orange juice he pours into a jug, he mixes 130 ml of pineapple juice.

Use your knowledge of equivalent ratios to complete the table.

orange juice (ml)	180		30		1620			45
pineapple juice (ml)		650		1170		26	5980	

Date: _____

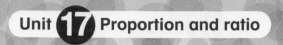

Number

Lesson 4: **Equivalent ratios (2)**

- Use equivalent ratios to calculate unknown amounts

1 Fill in the missing numbers so that the ratios in each column are equivalent.

1:3	4:1	6:5	9:7
2:6	8:2	12:☐	☐:21
3:☐	☐:4	☐:20	45:☐
☐:15	24:☐	42:☐	☐:63
8:☐	☐:10	☐:75	108:☐

2 Use the table to work out the answers to these equivalent ratio problems.

a To make a necklace, Liam uses 5 red beads for every 2 green beads.

i If he uses 15 red beads, how many green beads will he need? ☐

ii If he uses 12 green beads, how many red beads will he need? ☐

iii If he uses 65 red beads, how many green beads will he need? ☐

b To make a tahini dip, Phoebe uses 80 g of yoghurt for every 5 spoonfuls of tahini.

i If she uses 10 spoonfuls of tahini, how much yoghurt will she use? ▢

ii If she uses 320 g of yoghurt, how many spoonfuls of tahini will she use? ▢

iii If she uses 65 spoonfuls of tahini, how much yoghurt will she use? ▢

3 Use equivalent ratios to find the missing numbers in the table. Show your working out in the box.

> **Playdough recipe**
>
> $2\frac{1}{2}$ cups flour
>
> $\frac{3}{4}$ cup salt
>
> water
>
> food colouring
>
> $1\frac{1}{4}$ tablespoons of vegetable oil

flour (cups)	$2\frac{1}{2}$		$12\frac{1}{2}$	20	
salt (cups)	$\frac{3}{4}$	$2\frac{1}{4}$			13·5

Date: _____

Geometry and Measure

Lesson 1: **Quadrilaterals**

- Identify, describe and sketch quadrilaterals

1 Name each of these quadrilaterals.

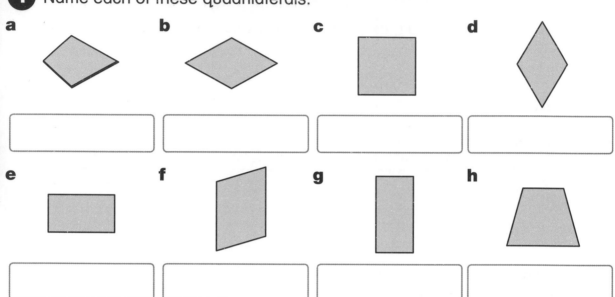

a   b   c   d

e   f   g   h

2 Write yes or no to answer each question. You may need to write a short explanation to classify and answer some of the questions.

Property	square	oblong	parallelogram	rhombus	kite	trapezium
opposite sides are equal						
opposite sides are parallel						
adjacent sides are equal						
all angles are 90°						
opposite angles are equal						
diagonals are of equal length						

Geometry and Measure

3 Draw the pairs of shapes. Mark the following:
- sides that are parallel or perpendicular, and of the same length
- similar angles and right angles
- diagonals, indicating whether they bisect or not.

Write two or three sentences to describe some of the similarities and differences between the shapes.

rectangle	rhombus

_____

_____

_____

square	trapezium

_____

_____

_____

4 Each description below lists the properties of one of the quadrilaterals named in the shaded box. Write the letter of the description below the name of the quadrilateral.

kite	rhombus	trapezium	oblong	square	parallelogram

A
- Opposite sides are equal and parallel
- All angles are right angles

B
- Two pairs of adjacent sides equal
- One pair of opposite angles is equal
- One diagonal bisects the other
- Diagonals intersect at right angles

C
- Opposite sides are parallel
- All sides are equal
- All angles are right angles

D
- Opposite sides are parallel
- All sides are equal
- Opposite angles are equal
- Diagonals bisect each other at right angles

E
- Opposite sides are equal and parallel
- Opposite angles are equal

F
- One pair of opposite sides is parallel

Date: _____

Geometry and Measure

Lesson 2: **Parts of a circle**

- Identify and label parts of a circle

You will need
- ruler

1 Draw a line from each label to the correct part of the circle.

radius

centre

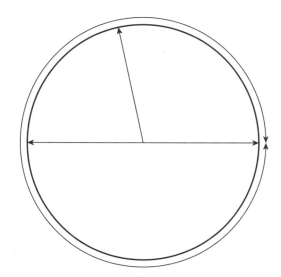

diameter

circumference

2 Mark and label a radius and a diameter on each circle.

a

b

c

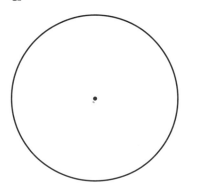

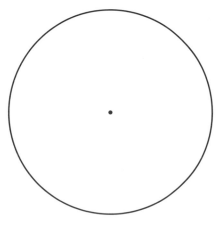

3 Write the missing numbers in each table. Show your working out.

a

Radius	Diameter
6 cm	
13 cm	
47 cm	
84 cm	
126 cm	
458 cm	

b

Diameter	Radius
36 cm	
82 cm	
98 cm	
146 cm	
258 cm	
534 cm	

4 What is the diameter of the large grey circle? ☐ cm. Show your working out.

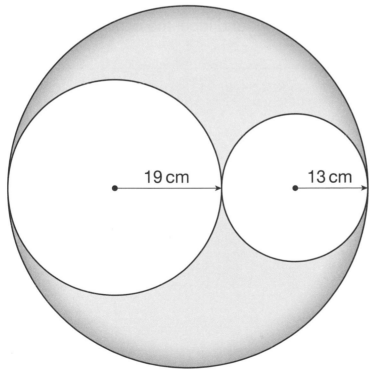

Working out

Date: _____

145

Geometry and Measure

Lesson 3: **Constructing circles**

- Construct circles of a given radius or diameter

You will need
- compass
- ruler
- cardboard (optional)
- tape (optional)

1 Mei draws some circles using a pencil and string method. She wants to change the size of the circle.

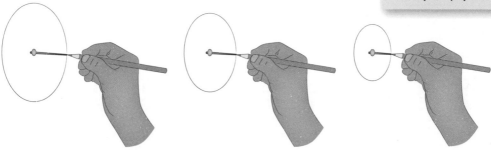

a How can she make a larger circle?

_____

b How can she make a smaller circle?

_____

2 Use a compass to draw a circle. Mark and label the centre, radius, diameter and circumference.

3 Draw a circle with a radius of 4 cm. Use the radius to calculate the diameter of the circle.

Diameter: [　　　] cm

4 Four radii (r) of a circle with centre C are marked on the coordinate grid. What are the coordinates of a point on the circumference of the circle with the:

a largest *y* coordinate?
(_____, _____)

b smallest *y* coordinate?
(_____, _____)

c largest *x* coordinate?
(_____, _____)

d smallest *x* coordinate?
(_____, _____)

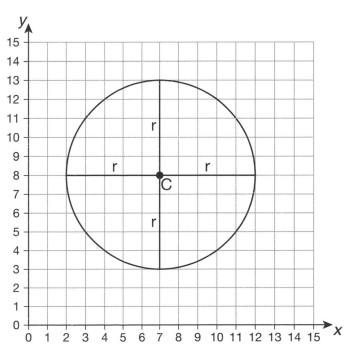

Date: _____

147

Geometry and Measure

Lesson 4: **Rotational symmetry**

• Identify rotational symmetry in familiar shapes, patterns or images

1 Place a tick (✓) in the boxes to indicate whether each shape has reflective symmetry, rotational symmetry or both.

a

b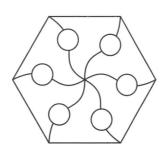

c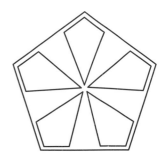

Reflective ☐   Reflective ☐   Reflective ☐

Rotational ☐   Rotational ☐   Rotational ☐

2 How many times can each shape be rotated around its centre so that it fits on top of itself?

a

b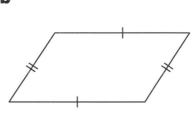

c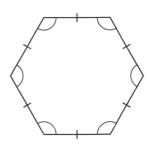

☐   ☐   ☐

d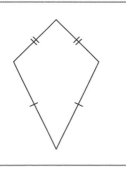

e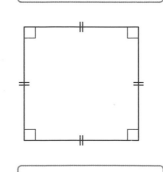

f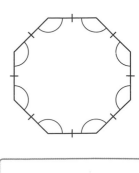

☐   ☐   ☐

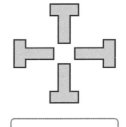

Geometry and Measure

3 How many times can each logo be rotated around its centre so that it fits on top of itself?

a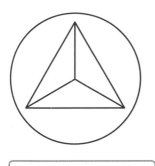

b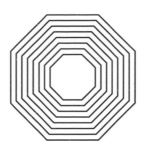

c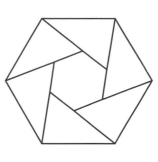

4 Write the order of rotational symmetry for each shape.

a

b

c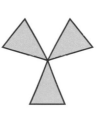

d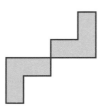

5 Complete the grids so that they have rotational symmetry about the centre.

a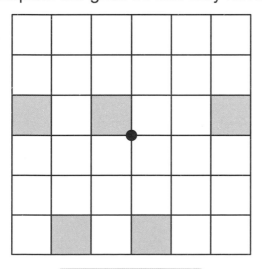

order 2

b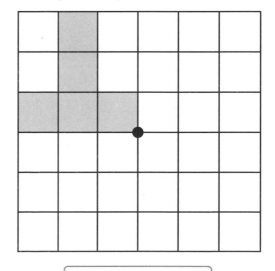

order 4

Date: _____

Lesson 1: **Identifying and describing compound 3D shapes**

- Identify and describe compound 3D shapes

1 Each compound 3D shape below is made of two component shapes. Name the component shapes.

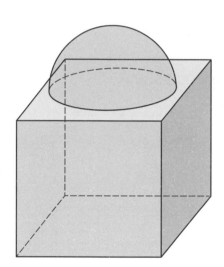 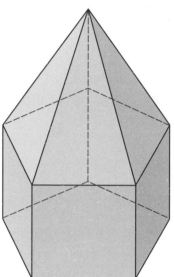

a Component shapes

b Component shapes

 Complete the table.

Compound shape	Faces	Vertices	Edges

Compound shape	Faces	Vertices	Edges
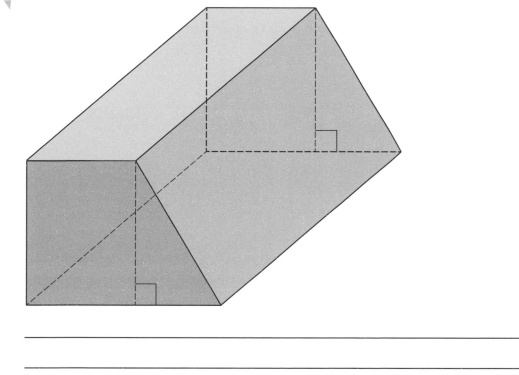			

3 Describe four properties of this compound shape. Number your properties 1 to 4.

_____

_____

_____

_____

Date: _____

Geometry and Measure

Geometry and Measure

Lesson 2: **Sketching compound 3D shapes**

• Sketch compound 3D shapes

1 Complete the sketch of a second cube.

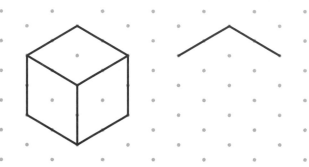

2 Copy the sketch of each shape. Try to include edges hidden from view.

cuboid

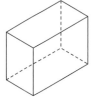

triangular prism

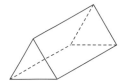

Geometry and Measure

3 Sketch the L shape.

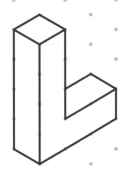

4 Choose two of the 3D letters to sketch.

Date: _____

Lesson 3: **Identifying nets**

- Identify the nets for different 3D shapes

1 Only one of these is the net for a cube. Draw a ring around the correct net.

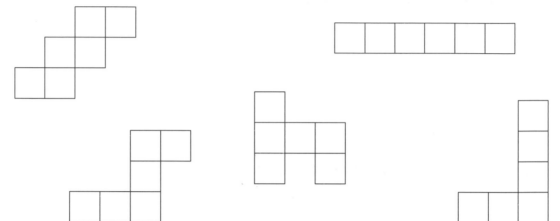

2 Look at the diagrams and answer the questions.

Example:

The square-based pyramid is made up of

one square and four triangles.

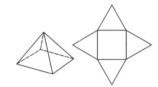

a The triangular pyramid (tetrahedron) is made up of

_____

_____

_____.

b The triangular prism is made up of

_____

_____

_____.

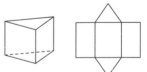

c The pentagonal prism is made up of

_____

_____

_____.

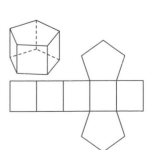

Geometry and Measure

3 Decide whether each of these nets will form a prism, a pyramid or a shape that is not solid. Write the letter of each shape in the correct part of the table.

Prism	Pyramid	Does not form a solid shape

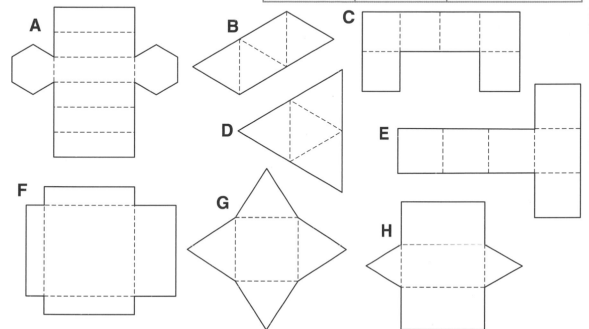

A   B   C   D   E   F   G   H

4 Which of the following nets will make a square-based pyramid? Tick Yes or No and explain your answer.

A 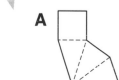   B 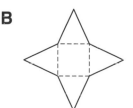   C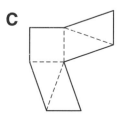

Net	Yes	No	Explanation
A			
B			
C			

Date: _____

Lesson 4: **Sketching nets**

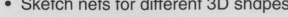

- Sketch nets for different 3D shapes

You will need
- triangular dot paper

1 Sketch the net for a cube.

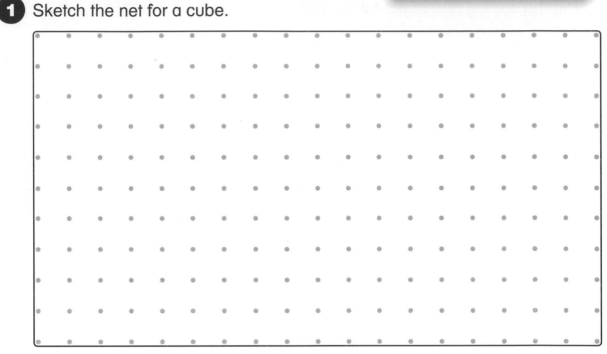

2 a Sketch the net for a square-based pyramid.

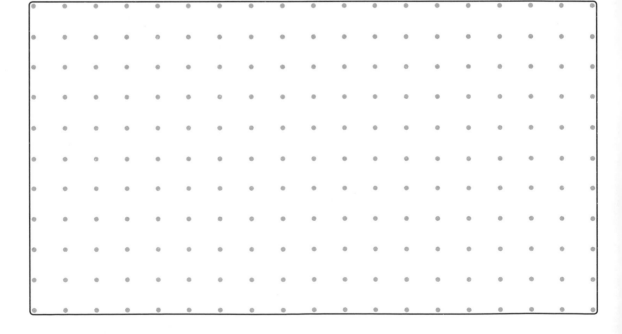

Geometry and Measure

Geometry and Measure

b Sketch the net for a triangular prism.

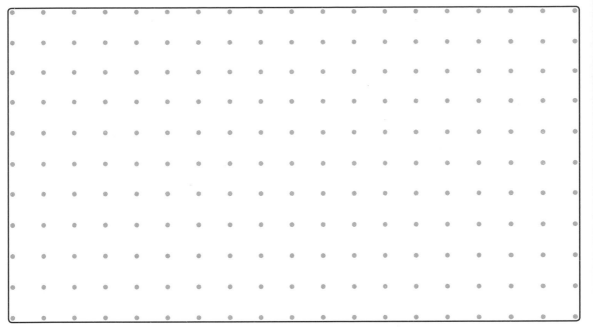

3 Grace has drawn nets for two 3D shapes. She thinks she has made some mistakes. Describe and critique the mistakes Grace has made.

a

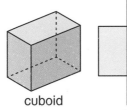

cuboid

net of the cuboid

b

pentagonal pyramid

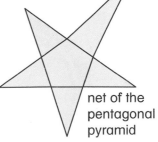

net of the pentagonal pyramid

_____

_____

_____

_____

_____

_____

4 Use triangular dot paper to sketch the net for the shape.

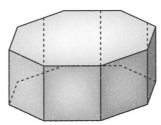

Date: _____

Geometry and Measure

Lesson 1: **Measuring angles**

• Classify, estimate and use a protractor to measure angles

1 Read the scale and write the measurement.

a [   ] °

b [   ] °

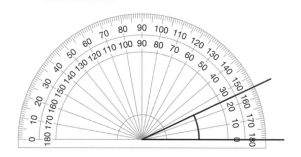

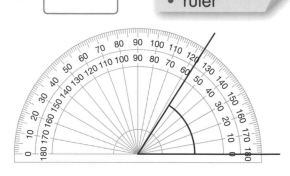

2 Estimate the size of each angle, then measure it using a protractor.

a [   ] ° Estimate: [   ] °

b [   ] ° Estimate: [   ] °

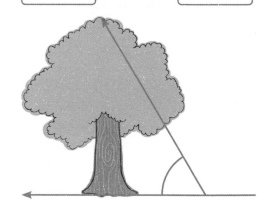

c [   ] ° Estimate: [   ] °

d [   ] ° Estimate: [   ] °

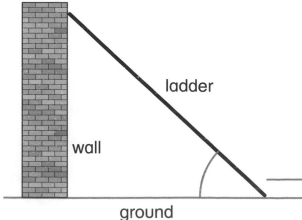

ladder

wall

ground

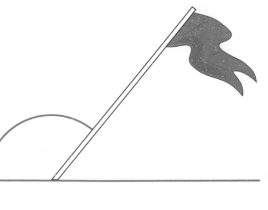

Geometry and Measure

3 Estimate the size of each angle, then measure it.

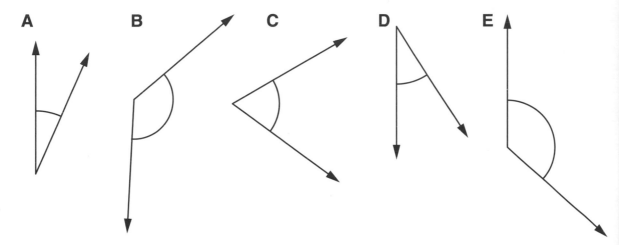

	A	B	C	D	E
Estimate:	°	°	°	°	°
Measurement:	°	°	°	°	°

4 Four games of Angle Golf have been played. Measure and write in the angles indicated. The angle that the ball hits the wall is the angle that it bounces off.

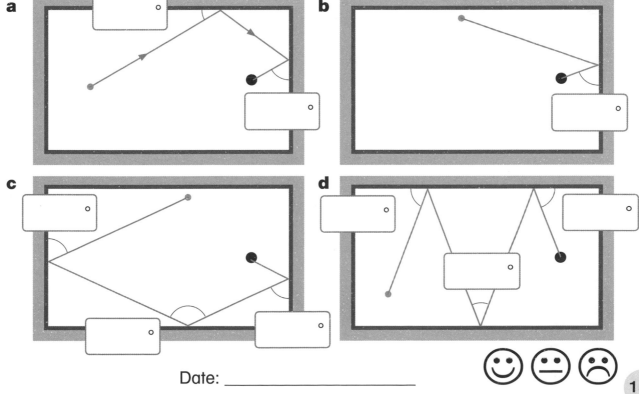

Date: _____

Lesson 2: **Drawing angles**

Geometry and Measure

• Use a protractor to draw angles

You will need
• protractor
• ruler

1 Use a ruler to draw the angle given.

a 40°

b 65°

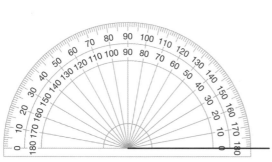

c 115°

d 160°

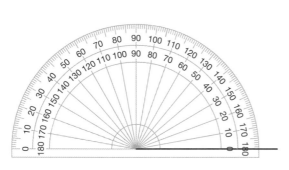

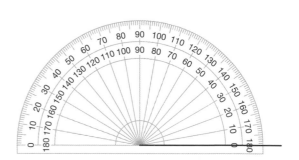

2 First estimate the position of the line to complete each angle. Draw it using a feint line. Then use a protractor and draw a solid line to complete the angle. How close was your estimate?

a 37°

b 83°

c 122°

d 176°

Geometry and Measure

3 A football stadium has been fitted with new floodlights. Three pairs project light at the following angles: 57°, 93° and 149°. Draw and label these angles on the picture. The first one, 57°, has been done for you.

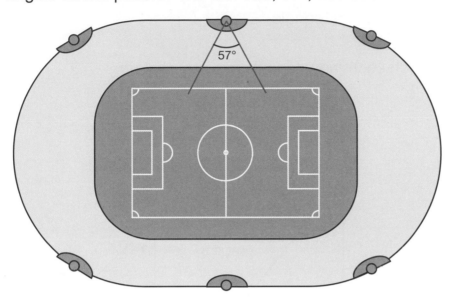

4 Play Angle Golf. Draw straight lines from the golf ball to the hole. Measure the angle between the path of the ball and the wall. The angle that the ball hits the wall is the angle that it bounces off.

Example:

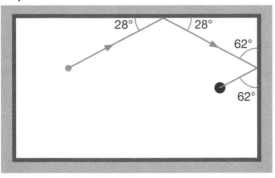

a

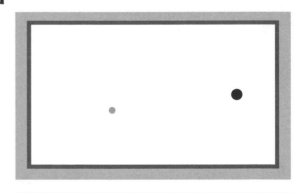

b

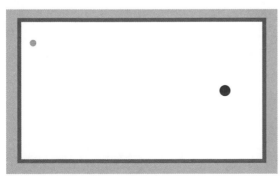

c

Date: _____

Lesson 3: **Calculating missing angles in a triangle (1)**

Geometry and Measure

- Calculate missing angles in a triangle

1 Draw lines to match each triangle to its missing angle.

| 39° | 46° | 55° |

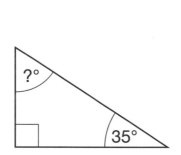

?°

35°

?°

51°

?°

44°

2 Write the missing angles.

a

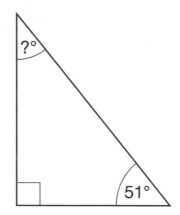

°

63°

49°

b

43°

°

c

21°

°

118°

d

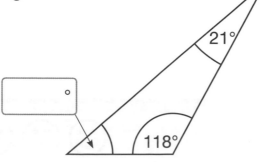

°

67°

°

3 The sum of angles in any triangle will always equal 180°. So, a shape cannot be a triangle if the angles don't add up to 180°.

Look at the angles of the shapes given in the table. Work out whether each shape can be a triangle or not. Tick the correct column.

Angles of shape	Is it a triangle?	
	Yes	No
37°, 48°, 95°		
84°, 39°, 57°		
55°, 68°, 58°		
29°, 53°, 98°		
17°, 25°, 137°		
26°, 66°, 89°		

4 Write the missing angles.

5

a

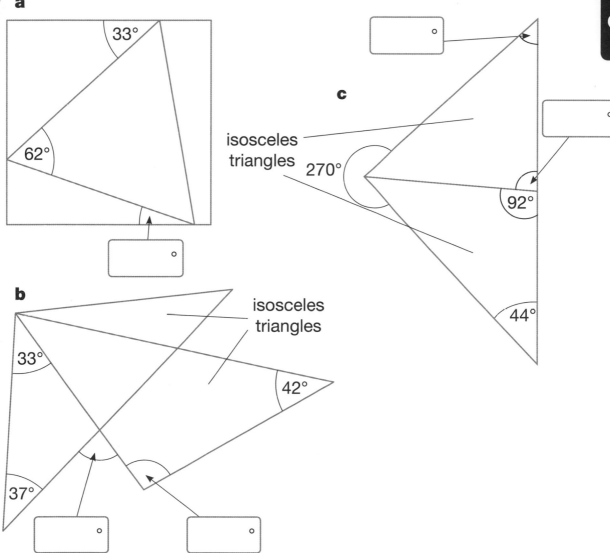

isosceles triangles

isosceles triangles

b

c

Date: _____

Geometry and Measure

Lesson 4: **Calculating missing angles in a triangle (2)**

- Calculate missing angles in a triangle

1 Identify each triangle based on its sides. Write **equilateral**, **isosceles** or **scalene**.

a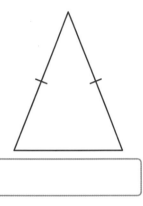

b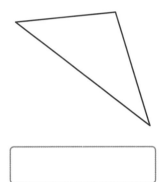

c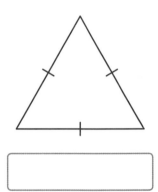

2 Shade the angles that are equal in each triangle.

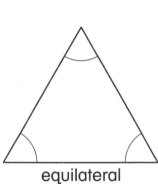

equilateral

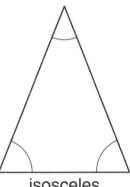

isosceles

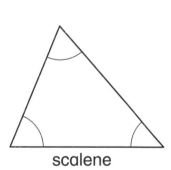

scalene

3 Find the missing angles. Show your working out in the boxes.

a = ____ °

b = ____ °

c = ____ °

Geometry and Measure

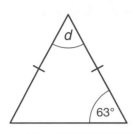

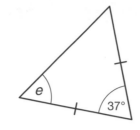

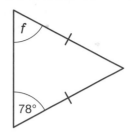

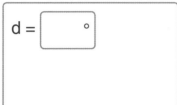

d = ⬚ °

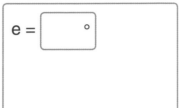

e = ⬚ °

f = ⬚ °

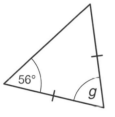

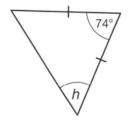

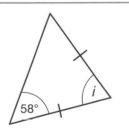

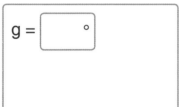

g = ⬚ °

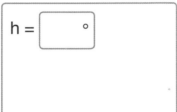

h = ⬚ °

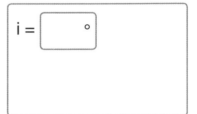

i = ⬚ °

4 Find the missing angles e, f, g and h.

5

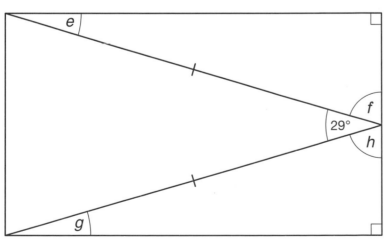

Working out

e = ⬚ °   f = ⬚ °   g = ⬚ °   h = ⬚ °

Date: _____

Lesson 1: **Converting time intervals (1)**

- Understand the relationship between units of time, and convert between them, including times expressed as a fraction or decimal

1 Convert the decimals to fractions or mixed numbers.

a 0·25 = [ ]

b 0·5 = [ ]

c 0·75 = [ ]

d 1·5 = [ ]

e 3·75 = [ ]

f 2·25 = [ ]

g 7·25 = [ ]

h 5·5 = [ ]

i 9·75 = [ ]

2 Convert the minutes to seconds.

a 1 min = [ ] s

b 3 min = [ ] s

c 5 min = [ ] s

d $\frac{1}{2}$ min = [ ] s

e $\frac{1}{4}$ min = [ ] s

f $\frac{3}{4}$ min = [ ] s

3 Convert the hours to minutes.

a 1 h = [ ] min

b 4 h = [ ] min

c 6 h = [ ] min

d $\frac{1}{2}$ h = [ ] min

e $\frac{1}{4}$ h = [ ] min

f $\frac{3}{4}$ h = [ ] min

4 Convert the minutes to minutes and seconds. Write your working in the boxes.

a $1\frac{1}{2}$ min = [ ] min [ ] s

b $3\frac{1}{4}$ min = [ ] min [ ] s

Geometry and Measure

Geometry and Measure

c 6·75 min = ▢ min ▢ s   **d** 8·5 min = ▢ min ▢ s

5 Convert the hours to hours and minutes. Write your working in the boxes.

a $2\frac{1}{2}$ h = ▢ h ▢ min   **b** $1\frac{3}{4}$ h = ▢ h ▢ min

c 4·25 h = ▢ h ▢ min   **d** 9·75 hours = ▢ h ▢ min

6 Answer these problems.

a Mi-Cha leaves home at 8:00 a.m. She gets to school 0·75 minutes before 8:10 a.m.

How many minutes and seconds was her walk to school?

▢ min ▢ s

b Adisa begins his homework at 6:05 p.m. He finishes it 0·25 hours before the clock turns 9:00 p.m.

How many hours and minutes did it take Adisa to complete

his homework? ▢ h ▢ min

Date: _____

Lesson 2: **Converting time intervals (2)**

- Understand the relationship between units of time, and convert between them, including times expressed as a fraction or decimal

1 Convert the decimals to fractions or mixed numbers.

a 0·1 = [   ]   **b** 0·3 = [   ]   **c** 0·7 = [   ]

d 2·4 = [   ]   **e** 6·6 = [   ]   **f** 9·2 = [   ]

g 11·8 = [   ]   **h** 16·9 = [   ]   **i** 8·5 = [   ]

2 Convert between the units. Write each answer as a decimal.

a $\frac{1}{10}$ min = [   ] s   **b** $\frac{1}{10}$ h = [   ] min   **c** $\frac{5}{10}$ min = [   ] s

d $\frac{5}{10}$ h = [   ] min   **e** $1\frac{1}{2}$ min = [   ] s   **f** $2\frac{3}{4}$ h = [   ] min

3 Convert the minutes to minutes and seconds. Show your working out in the boxes.

a 4·1 min = [   ] min [   ] s   **b** 2·3 min = [   ] min [   ] s

c 5·6 min = [   ] min [   ] s   **d** 8·8 min = [   ] min [   ] s

Geometry and Measure

4 Convert the hours to hours and minutes. Show your working out in the boxes.

a $\frac{41}{10}$ h = ☐ h ☐ min

b $\frac{63}{10}$ h = ☐ h ☐ min

c $\frac{117}{10}$ h = ☐ h ☐ min

d $\frac{149}{10}$ h = ☐ h ☐ min

Geometry and Measure

5 Answer these problems. Show your working out in the boxes.

a Tom leaves to go to work at 7:30 a.m. He gets to his office at 0·3 minutes before 7:55 a.m.

How many minutes and seconds was his journey to work?

☐ min ☐ s

b Florence begins an art project at 6:10 p.m. She finishes it 0·3 hours before the clock turns 9:52 p.m.

How many hours and minutes does it take Florence to complete?

her project? ☐ h ☐ min

Date: _____

Geometry and Measure

Lesson 3: **Capacity and volume (1)**

• Understand the difference between volume and capacity

1 Think about each statement and the container(s) it describes. Write the letter of the container(s) next to each statement.

A

B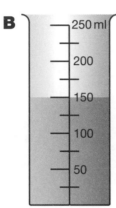

C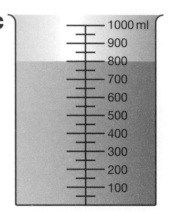

Statement	Container
The capacity is 1000 ml.	
The volume of liquid in the container is 150 ml.	
The capacity is 500 ml.	
The volume of liquid in the container is 500 ml.	
The capacity is 250 ml.	
The volume of liquid in the container is 800 ml.	

2 Write the capacity of each container and the volume of water it holds.

a

Capacity: ☐ ml

Volume: ☐ ml

b

Capacity: ☐ ml

Volume: ☐ ml

c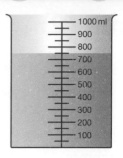

Capacity: ☐ ml

Volume: ☐ ml

d

Capacity: ☐ ml

Volume: ☐ ml

 Shade each container to show the amount of water. Then complete the scale.

a Capacity is 400 ml. Jug is half full.

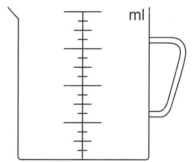

b Capacity is 200 ml. Volume of water in the jug is 75 ml.

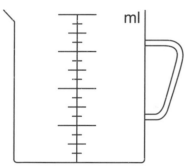

4 Explain the difference between volume and capacity. Use the diagrams to help you, adding detail to them where required.

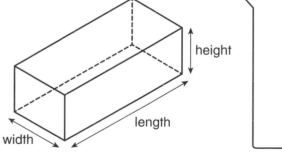

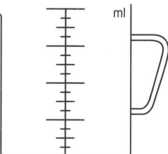

_____

_____

_____

Date: _____

Lesson 4: **Capacity and volume (2)**

- Understand the difference between volume and capacity

You will need
- paper for working out

Geometry and Measure

1 Work out the total capacity in litres of each set of plastic containers. Use paper to work out the answers.

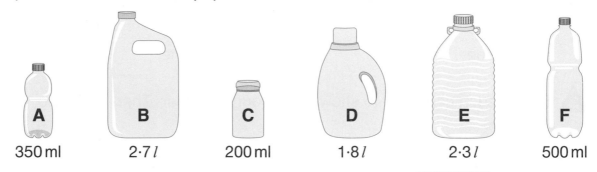

A	B	C	D	E	F
350 ml	2·7 l	200 ml	1·8 l	2·3 l	500 ml

a 1 container A, 1 container B and 1 container C = [ ] l

b 1 container D, 1 container E and 1 container F = [ ] l

c 2 container A, 2 container D and 1 container F = [ ] l

d 2 container B, 3 container C and 1 container E = [ ] l

2 Work out the total capacity in litres of each set of containers. Use paper to work out the answers.

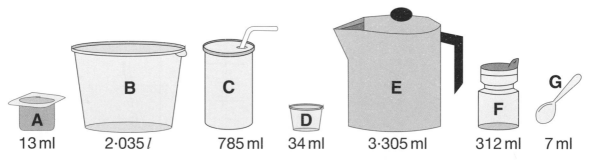

A	B	C	D	E	F	G
13 ml	2·035 l	785 ml	34 ml	3·305 ml	312 ml	7 ml

a 2 container A, 2 container B and 2 container C = [ ] l

b 4 container D, 2 container F and 13 container G = [ ] l

c 2 container E and 5 container F = [ ] l

d 77 containers A, 10 container B and 27 container D = [ ] l

3 Work out the answers to these problems. Show your working out in the boxes.

a Katie makes a 500 ml jug of rice pudding. How much rice pudding is left over if she pours 227 ml into one bowl and 0·198 *l* into another?

ml

b i If a bottle holds 175 ml of fizzy soda water, how much will 5 bottles

hold? *l*

ii How many bottles would I need to make 1·75 *l*?

4 Work out the answers to these problems. Show your working out in the boxes.

a The Jones family used 79·8 litres of water on Monday, 87 *l* 146 ml on Tuesday and 123·67 ml on Wednesday. How much water did they

use over the three days? *l*

b Melissa bought 13 cans of cola and 16 cans of lemonade. If the capacity of the cola cans is 330 ml and the lemonade cans is 0·47 *l*,

what is the combined capacity of all the cans? *l*

Date: _____

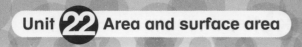

Lesson 1: **Calculating the area of a triangle (1)**

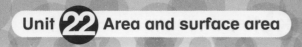

Geometry and Measure

• Prove that the area of a right-angled triangle is half the area of its related rectangle

You will need
• ruler

1 Split each rectangle into two triangles. Each triangle must be half the area of the rectangle.

2 Draw two rectangles on the squared paper with different dimensions that have an area of 18 square units. Mark each rectangle to create a triangle with an area of 9 square units.

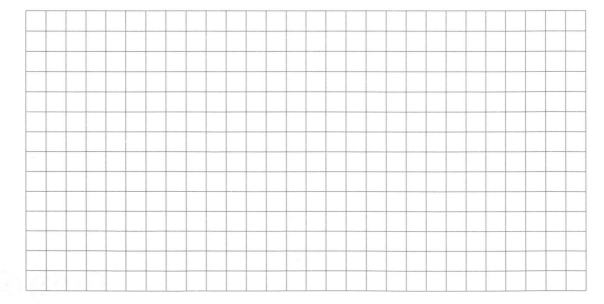

3 Work out the area of each triangle by drawing a related rectangle.

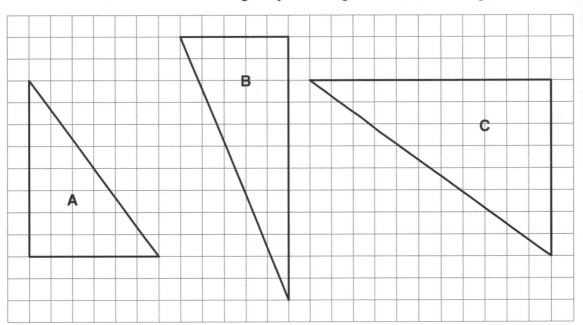

Area A:

 square units

Area B:

[ ] square units

Area C:

[ ] square units

4 Work out the area of each triangle.

a

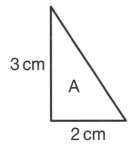

Area A:

[ ] cm².

b

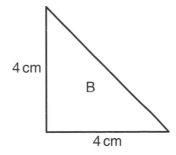

Area B:

[ ] cm².

c

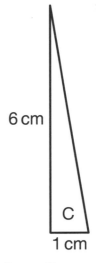

Area C:

[ ] cm².

Date: _____

Lesson 2: **Calculating the area of a triangle (2)**

- Use the area of a related rectangle to find the area of a right-angled triangle

You will need
- paper for working out (optional)

Geometry and Measure

1 Each rectangle is split into two identical right-angled triangles. Calculate the area of each triangle.

a
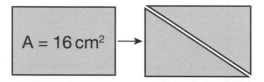

A = 16 cm²

Area of each triangle = ☐ cm²

b
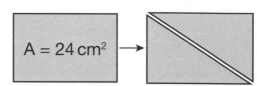

A = 24 cm²

Area of each triangle = ☐ cm²

c

A = 50 cm²

Area of each triangle = ☐ cm²

d

A = 76 cm²

Area of each triangle = ☐ cm²

2 Calculate the area of each triangle.

a
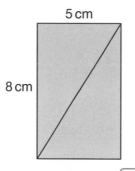

5 cm

8 cm

Area of each triangle = ☐ cm²

b

14 m

4 m

Area of each triangle = ☐ m²

c

15 cm

12 cm

Area of each triangle = ☐ cm²

d
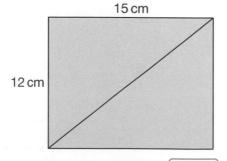

3 m

48 m

Area of each triangle = ☐ m²

Geometry and Measure

3 Calculate the area of each triangle.

a
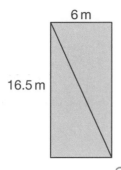
6 m

16.5 m

Area of each triangle = ☐ m²

b
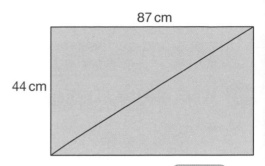
87 cm

44 cm

Area of each triangle = ☐ cm²

c
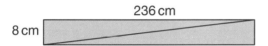
236 cm

8 cm

Area of each triangle = ☐ cm²

d
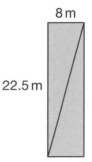
8 m

22.5 m

Area of each triangle = ☐ m²

4 Work out the missing side lengths x and y.

a
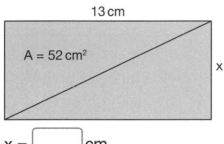
13 cm

A = 52 cm²

x

x = ☐ cm

b
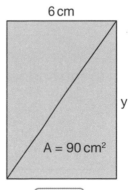
6 cm

y

A = 90 cm²

y = ☐ cm

Date: _____

Lesson 3: **Surface area (1)**

- Understand the relationship between area of 2D shapes and surface area of 3D shapes

You will need
- coloured pencils

Geometry and Measure

1 What are the 2D shapes that cover the surface of each 3D shape?

a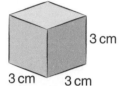
3 cm
3 cm   3 cm

_____

b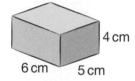
4 cm
6 cm   5 cm

_____

c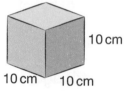
10 cm
10 cm   10 cm

_____

d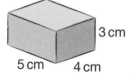
3 cm
5 cm   4 cm

_____

2 Look at each of the shapes in **1**. How would you find the surface area of the shape?

a _____

_____

b _____

_____

c _____

_____

d _____

_____

Geometry and Measure

3 For each 3D shape, label the faces and then write how you would work out the surface area of the shape. The first one has been done for you.

a

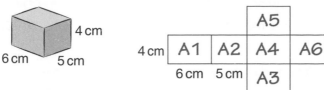

Surface area = | A1 + A2 + A3 + A4 + A5 + A6 |

b

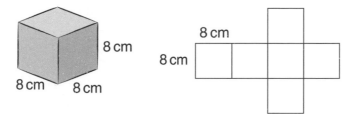

Surface area =

c

Surface area =

d

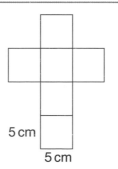

4 Look at each of nets in **3**. For each net, use a different coloured pencil to shade faces with the same area.

Date: _____

Lesson 4: **Surface area (2)**

- Understand the relationship between area of 2D shapes and surface area of 3D shapes

You will need
- coloured pencils

1 Look at each 3D shape. What are the different 2D shapes that cover the surface of each 3D shape? How many of each 2D shape are there?

a

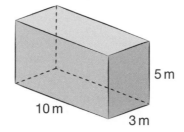

5 m
10 m
3 m

b

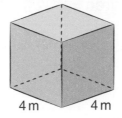

4 m   4 m

_____

_____

_____

_____

2 For each net, use a different coloured pencil to shade faces with the same area. The first one is done for you.

a

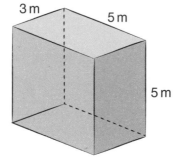

3 m   5 m

5 m

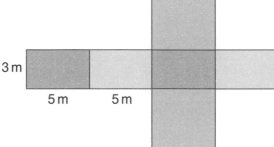

3 m

5 m   5 m

b

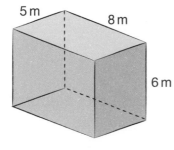

5 m   8 m

6 m

5 m

8 m   6 m

c

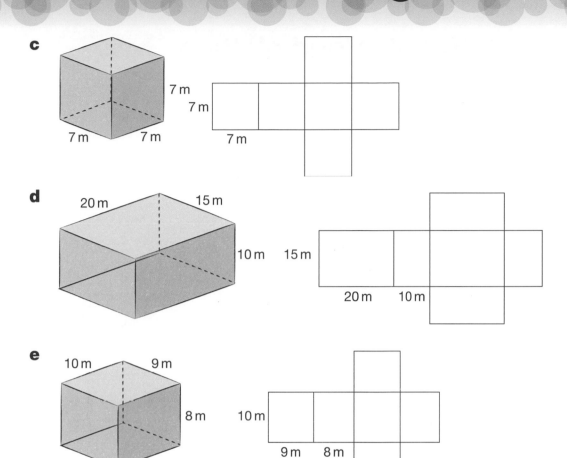

d

e

3 What conjectures can you make about finding the surface area of a cube?

_____

_____

_____

_____

4 What conjectures can you make about finding the surface area of a cube?

_____

_____

_____

_____

Date: _____

Lesson 1: **Reading and plotting coordinates (1)**

Geometry and Measure

• Read coordinates in all four quadrants

1 Write the coordinates of each point. The points are all in the first quadrant.

Point	Coordinates
A	(____, ____)
B	(____, ____)
C	(____, ____)
D	(____, ____)

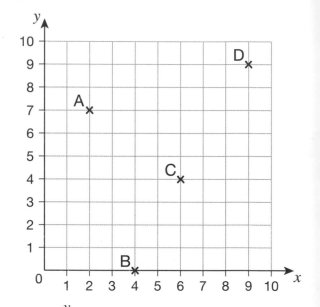

2 Write the coordinates of each point. The points are positioned across all four quadrants.

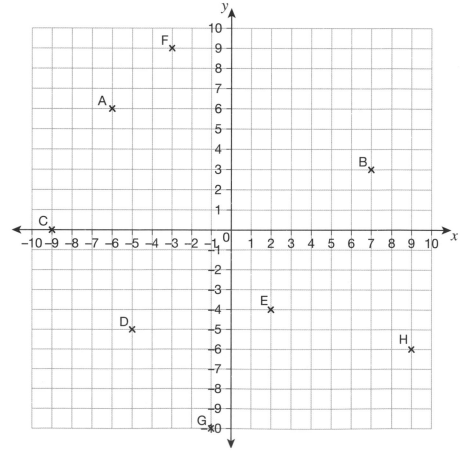

Geometry and Measure

Point	Coordinates
A	(___, ___)
B	(___, ___)
C	(___, ___)
D	(___, ___)

Point	Coordinates
E	(___, ___)
F	(___, ___)
G	(___, ___)
H	(___, ___)

3 Write the coordinates of the points in the form given.

Point	Coordinates (decimals)
P	(___, ___)
Q	(___, ___)

Point	Coordinates (fractions)
R	(___, ___)
S	(___, ___)

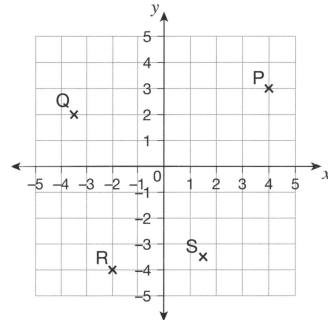

4 Identify the coordinates of each point.

a A point P, 5 units left of point Q (7, 2) P: (___, ___)

b A point M, 6 units up from point N (–4, –3) M: (___, ___)

c A point E, 7 units right of point F (–6, 0) E: (___, ___)

d A point Z, 8 units down from point Y (10, 4) Z: (___, ___)

e A point A, 3 units left and 4 units up from point B (5, –1) A: (___, ___)

f A point C, 9 units right and 8 units down from point D (–2, –2)
C: (___, ___)

g A point X, 7 units left and 9 units up from point Y (–1, –4) X: (___, ___)

h A point G, 8 units right and 7 units down from point
H (–7, 3) G: (___, ___)

Date: _____

Lesson 2: **Reading and plotting coordinates (2)**

Geometry and Measure

• Plot coordinates in all four quadrants

1 Plot and label the points in the first quadrant.

Point	Coordinates
A	(1, 4)
B	(3, 3)
C	(0, 9)
D	(8, 6)
E	(10, 9)
F	(4, 0)
G	(7, 5)
H	(2, 7)

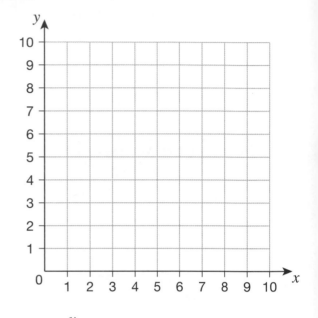

2 Plot and label the points in all four quadrants.

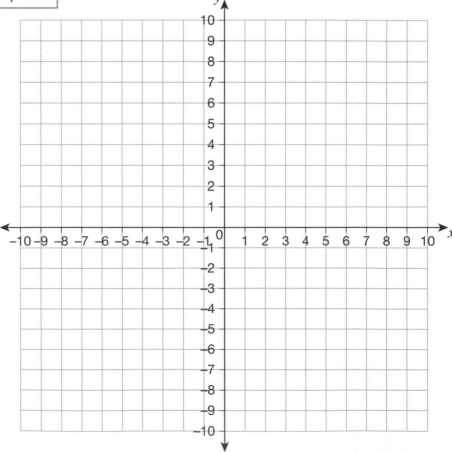

Geometry and Measure

Point	Coordinates
A	(−1, 4)
B	(3, −6)
C	(−6, −9)
D	(−4, 8)

Point	Coordinates
E	(−5, −5)
F	(7, 7)
G	(−2, 10)
H	(2, −1)

 Plot and label the points in all four quadrants.

Point	Coordinates
A	(−2, 2·5)
B	(1·25, −4)
C	(−3½, −3½)
D	(2¼, 5)
E	(0, 4¾)
F	(−3·75, 2)
G	(−1·5, −1·25)
H	(2¼, −4¼)

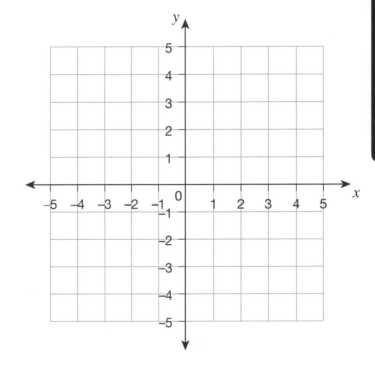

4 Use the clues to find where to draw each object on the map.

a A shark at (−6, 6).

b A boat at an x coordinate −3 and a y coordinate −2 from the shark.

c A lighthouse at (2, −4·5).

d A hut at (−3, 4½).

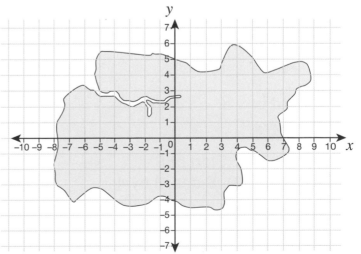

Date: _____

Geometry and Measure

Lesson 3: **Plotting lines and shapes across all four quadrants (1)**

* Plot points to form shapes in all four quadrants

You will need
* ruler

1 Plot and label each set of points.

Join the points to form two squares.

ABCD: (–4, 4) (1, 4) (1, –1) (–4, –1)

EFGH: (–1, –5) (3, –5) (3, –9) (–1, –9)

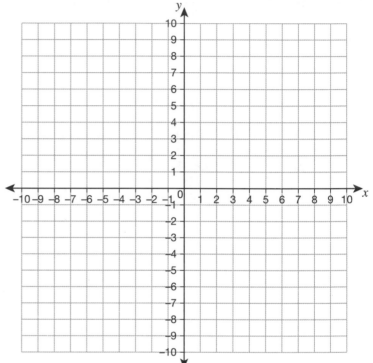

2 Draw each rectangle described. Write the coordinates of the vertices. There is more than one answer.

a JKLM: Start at J (–7, 8). Sides: 6 by 3.

Coordinates of each vertex:

K: (____, ____)

L: (____, ____)

M: (____, ____)

b PQRS: Start at P (–2, –3). Sides: 8 by 4.

Coordinates of each vertex:

Q: (____, ____)

R: (____, ____)

S: (____, ____)

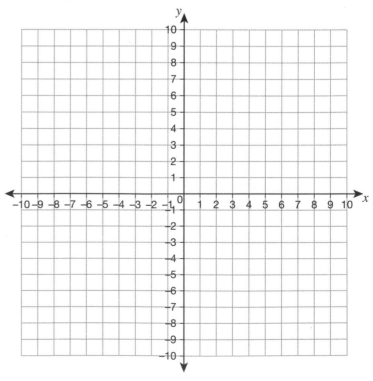

Geometry and Measure

3 Draw each shape and identify the coordinates of the missing vertex.

a ABC is an isosceles triangle.

The vertices of the base of triangle AB are: A (−2, −4) B (6, −4).

The triangle has a height of 4 units.

What are the coordinates of vertex C? (___, ___)

b EFGH is a parallelogram.

E: (−5, 9) F: (8, 9) G: (4, 3)

What are the coordinates of vertex H? (___, ___)

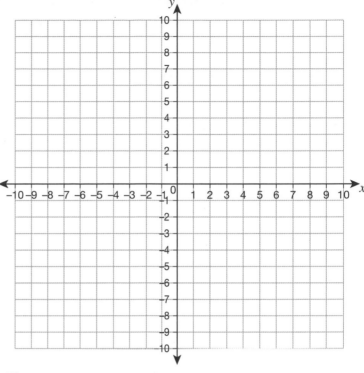

4 The position of 10 houses can be plotted on a map.

Each house is the same distance from its next door neighboor.

Four of the houses have been plotted on the map.

Find the positions of the other 6 houses. Plot and join these points on the map.

Don't forget that the diagonal distance across a square is not the same as a side distance.

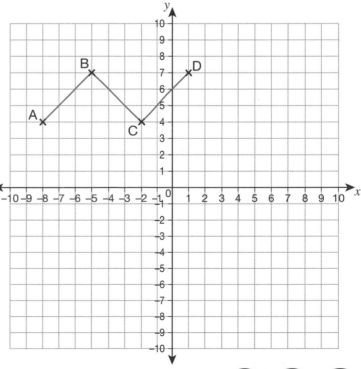

Date: _____

Lesson 4: **Plotting lines and shapes across all four quadrants (2)**

- Plot points to form lines in all four quadrants

You will need
- ruler

1 Each pair of points lie on a different line.

Plot and join the points with a straight line that extends to the edges of the coordinate grid.

a A: (8, 3) B: (−6, −6)

b C: (−7, 5) D: (4, 4)

c E: (−9, −4) F: (1, 8)

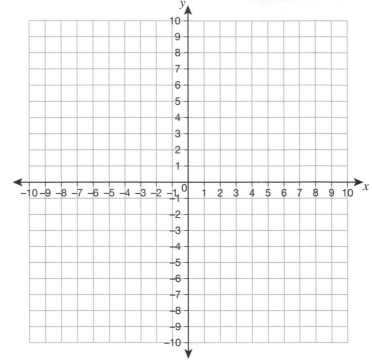

2 Each pair of points lie on a different line.

Plot and join the points with a straight line that extends to the edges of the coordinate grid.

Write and label the positions of two other points that lie on each line.

a A: (4, 8) B: (−7, −3)

Point: ____: (____, ____)

Point: ____: (____, ____)

b C: (−6, 7) D: (2, −9)

Point: ____: (____, ____)

Point: ____: (____, ____)

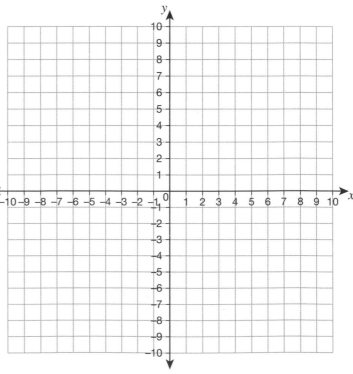

3 Each pair of points lie on a different line.

Plot and join the points with a straight line that extends to the edges of the coordinate grid.

Complete the missing coordinates.

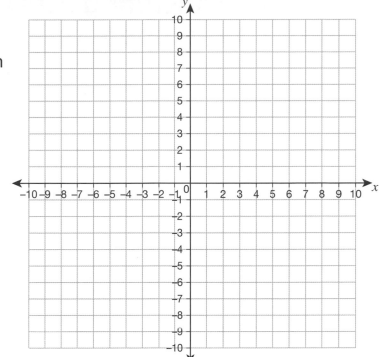

a A: (9, 2) B: (−9, −4)

Point: M: (_____, 0)

Point: N: (0, _____)

b C: (−7, 9) D: (2, −9)

Point: P: (_____, 0)

Point: R: (0, _____)

4

a Plot a line perpendicular to the line that joins the points A (−8, −3) and B (5, −8).

Identify two points that are on this line.

_____

_____

_____

_____

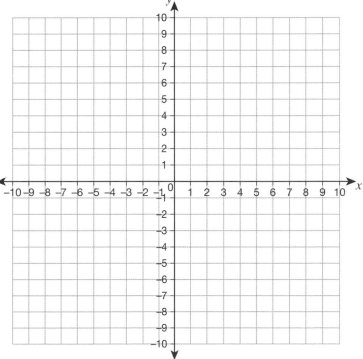

b Plot a line parallel to the line that joins the points C (−4, −6) and D (8, 4).

Identify two points that are on this line.

_____

Date: _____

Geometry and Measure

189

Lesson 1: **Translating 2D shapes on coordinate grids**

Geometry and Measure

• Translate shapes across all four quadrants on a coordinate grid

You will need
• ruler

1 Follow the instructions to translate the shapes.

a Move Shape A
6 units left.

b Move Shape B
6 units right,
3 units up.

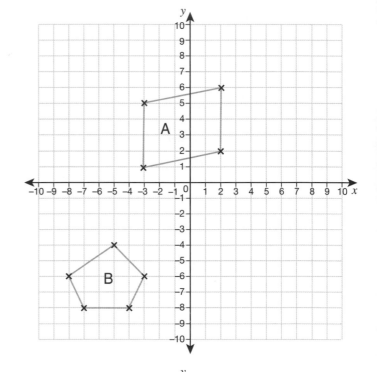

2 Translate the trapezium:
6 units right, 7 units up.

Translate the image:
7 units left, 6 units up.

Which single translation is
equivalent to these
two translations?

_____

_____

_____

_____

_____

_____

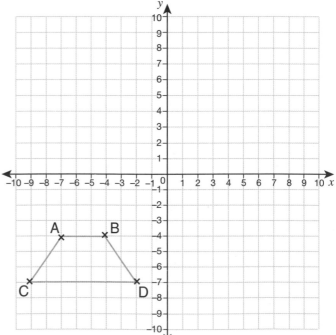

3 Use the translation shown in the grid to complete the table. Write the coordinates of the vertex A and its images. Find vertex A''' by following the pattern.

Vertex	*x*-coordinate	*y*-coordinate
A		
A'		
A''		
A'''		

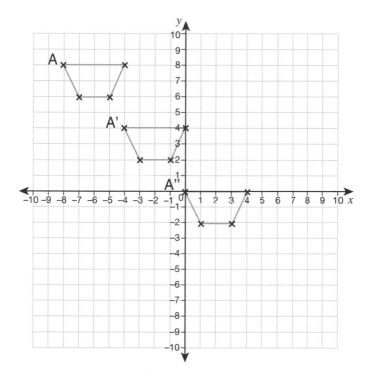

Geometry and Measure

4 Answer the problems.

a Car A moves from (6, 3) to (4, 1). Car B starts at (3, 5). Where will it move to if it uses the same translation as Car A?

(____, ____)

b Chess piece A moves from (1, –6) to (–3,–3). Chess piece B starts at (–5, –2). Where will it move to if it follows the same translation as A?

(____, ____)

Date: _____

Lesson 2: **Reflecting 2D shapes in a mirror line (1)**

Geometry and Measure

- Reflect shapes in horizontal and vertical mirror lines

1 Reflect each point in the mirror line. Label the images A'–F'.

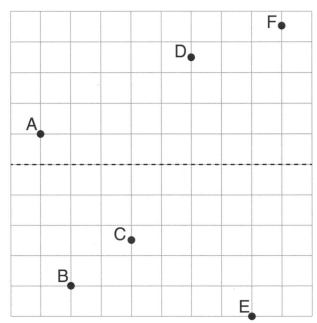

2 Reflect each shape in the mirror line.

a

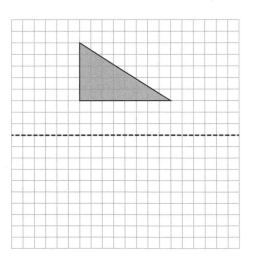

b

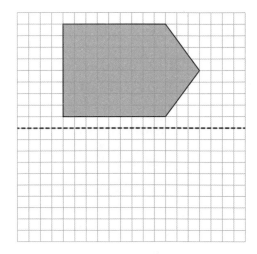

Geometry and Measure

c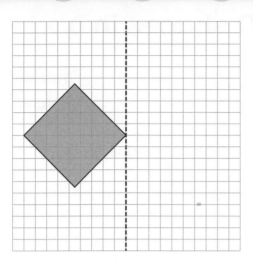

d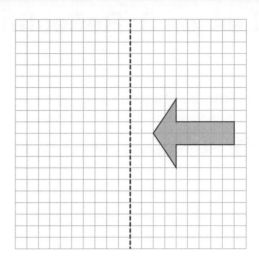

3 Reflect each shape in the mirror line.

a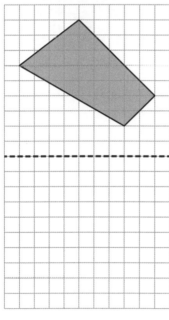

b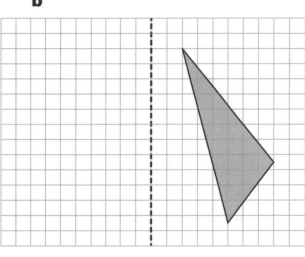

4 Reflect each shape in the mirror line.

a 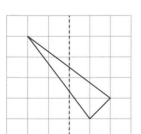 **b** 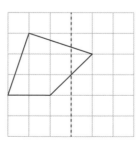 **c** 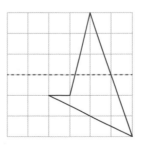 **d**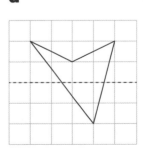

Date: _____

Lesson 3: **Reflecting 2D shapes in a mirror line (2)**

- Reflect shapes in diagonal mirror lines

You will need
- ruler

1 Reflect each point in the mirror line. Label the images A'–F'.

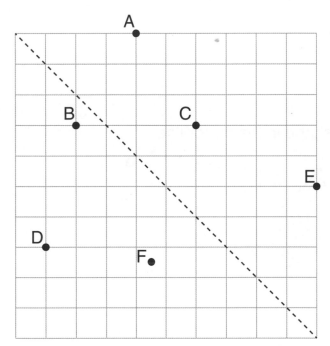

2 Reflect each shape in the mirror line.

a

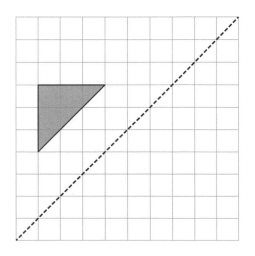

b

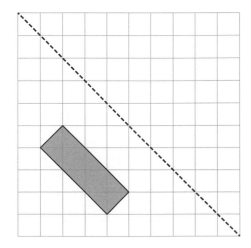

Geometry and Measure

Geometry and Measure

c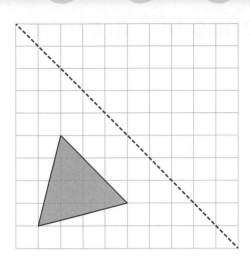

d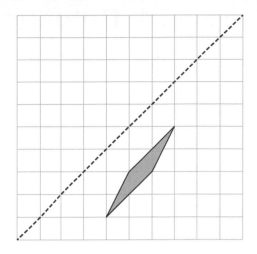

3 Reflect each shape in the mirror line.

a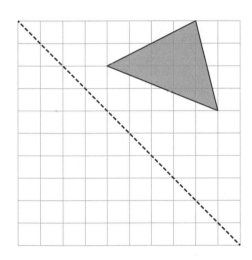

b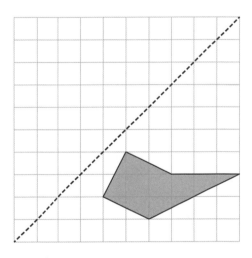

4 Reflect each shape in the mirror line.

a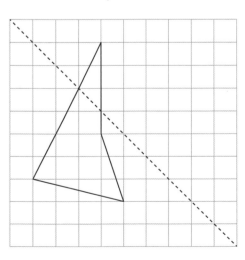

b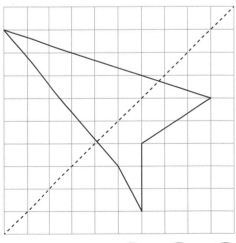

Date: _____

195

Lesson 4: **Rotating shapes 90° around a vertex**

- Rotate shapes 90° around a vertex (clockwise or anticlockwise)

You will need
- tracing paper
- ruler

1 Trace the shape and place a dot on the point of rotation, vertex A.

Place the tip of a pencil on vertex A and turn the tracing paper the indicated direction and amount.

Draw the image of the shape to complete the drawing.

Rotate the arrow 90° clockwise about vertex A.

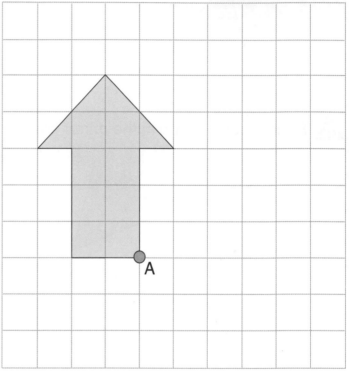

2 Draw the image of the triangle under a rotation of 90° clockwise about vertex A.

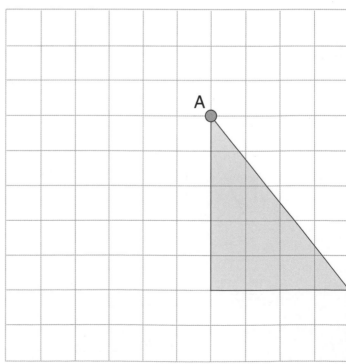

3 Draw the image of the trapezium under a rotation of 90° anticlockwise about vertex A.

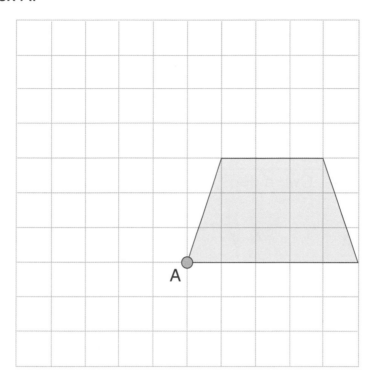

4 Complete the two-step transformation.

Step one: translation: 4 right and 5 down

Step two: rotation: 90° clockwise about vertex A.

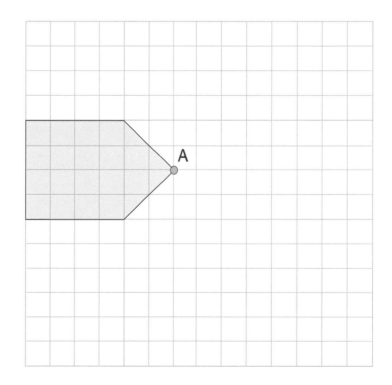

Geometry and Measure

Date: _____

Lesson 1: **Venn and Carroll diagrams**

- Record, organise, represent and interpret data in Venn and Carroll diagrams

Statistics and Probability

1 Class 6 investigated the statistical question: 'Do more learners own a dog than a cat?'

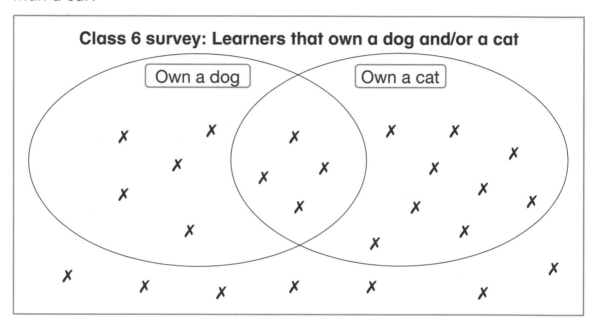

Class 6 survey: Learners that own a dog and/or a cat

a How many learners in Class 6 own:

i a cat?

ii a dog?

iii both?

iv neither?

b Does the data help to answer the statistical question?

Explain how.

_____

_____

_____

_____

Statistics and Probability

2 Leona predicts that owning a car means a person is less likely to own a bicycle. She conducts a survey that asks 36 adults: 'Do you own a car? Do you own a bicycle?'

a Use the results to complete the Carroll diagram.

Number of adults who own a car	Number of adults who own a bike
⊦⊦⊦ ⊦⊦⊦ ⊦⊦⊦ ////	⊦⊦⊦ ⊦⊦⊦ /
Number of adults who own both a car and a bike	**Number of adults who own neither a car nor a bike**
⊦⊦⊦	/

	Own a car	Do not own a car
Own a bike		
Do not own a bike		

b Write two conclusions you can draw from the data.

i _____

_____

ii _____

_____

c Does the data agree with Leona's prediction? [   ]

Explain your answer.

_____

_____

_____

_____

Statistics and Probability

3 Conduct an investigation similar to **2**.

a Write your statistical question here:

_____

b What is your prediction?

_____

You will need to decide on the two choices that a person answering the question can make.

c Gather your data and present it here in a frequency diagram.

d Now use the data to complete a Carroll diagram.

e What conclusions can you draw from the data?

_____

_____

_____

f Does the data agree with your prediction? [   ] How do you know?

_____

_____

😊 😐 ☹️

Date: _____

Lesson 2: **Tally charts, frequency tables and bar charts**

* Record, organise, represent and interpret data in tally charts, frequency tables and bar charts

You will need

* spinner from Resource sheet 28: Spinners (4)
* pencil and paper clip, for the spinner
* coloured pencil

1 Maisie used a spinner labelled 3, 4, 4, 5, 6, 6. The chart shows the tally marks for the numbers Maisie spun. Complete the tally chart and use the data to construct a bar chart.

Numbers spun with a 3, 4, 4, 5, 6, 6 spinner

Number spun	Tally	Frequency
3	⫽⫽⫽ ⫽⫽	
4	⫽⫽⫽ ⫽⫽⫽ ⫽⫽⫽⫽	
5	⫽⫽⫽	
6	⫽⫽⫽ ⫽⫽⫽ ⫽⫽	

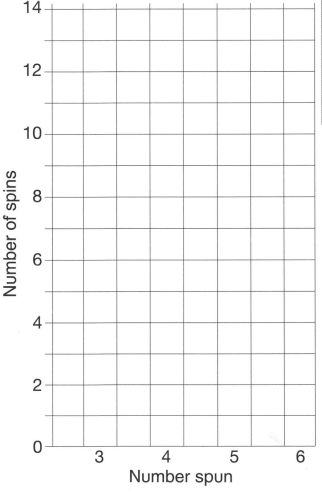

Numbers spun with a 3, 4, 4, 5, 6, 6 spinner

2 Write two conclusions you can draw from the data.

i _____

_____

_____

_____

ii _____

_____

_____

_____

Statistics and Probability

Statistics and Probability

3 **a** Work with a partner. One learner spins the spinner 50 times.
The other learner records each number spun, using a tally mark.

b Which number do you predict will be spun most frequently and why?

Numbers spun with a 3, 4, 4, 5, 6, 6 spinner		
Number spun	**Tally**	**Frequency**
3		
4		
5		
6		

Numbers spun by a 3, 4, 4, 5, 6, 6 spinner

Number of spins vs Number spun (bar chart, y-axis 0–20, x-axis 3 4 5 6)

c Complete the frequency column in the chart.

d Draw a bar chart of the data from the tally chart.

e How many times did you:

i spin a 4? ▢

ii spin a number greater than 4? ▢

iii spin a number less than 5? ▢

f Was your prediction correct? ▢ If not, can you explain why?

---

4 Mandisa wants to know how much water a typical household uses.
She predicts that the number is around 100 litres per day. She looks on
the internet and finds data for 50 households (litres per day):

103	128	132	164	123	105	128	174	140	171
174	167	149	146	159	106	179	145	176	179
160	106	122	164	136	111	106	137	153	159
135	146	153	158	151	145	171	116	165	149
120	115	178	117	108	146	120	133	120	177

Statistics and Probability

a Complete a grouped tally chart and frequency table for the given data. Decide on the most appropriate interval to use.

Water use (litres per day)	Tally	Frequency
101–110		
111–120		

b Complete a bar graph to show the results.

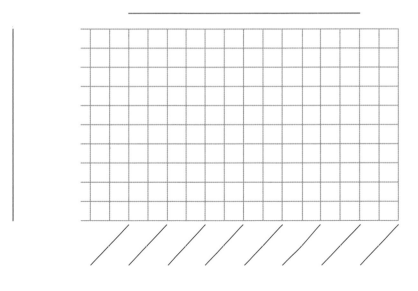

c Was Mandisa's prediction correct? ☐ How do you know?

_____

d Write two other conclusions you can draw from the data.

i _____

ii _____

Date: _____

Statistics and Probability

Lesson 3: **Waffle diagrams and pie charts**

- Record, organise, represent and interpret data in waffle diagrams and pie charts

You will need

- Resource sheet 14: Waffle diagrams
- coloured pencils
- ruler

1 Convert the frequency values to percentages.

Number of trees counted	Frequency	Percentage
oak	7	
birch	6	
pine	2	
beech	5	

2 Ten learners in a class were asked the question: 'What is your favourite flavour of fruit squash?'

Alicia predicts that there will be twice as many learners who prefer blackcurrant than lemon.

a Complete the frequency and percentage columns in the chart.

Juice	Tally	Frequency	Percentage
blackcurrant	ⵜⵜⵜ		
apple	/		
strawberry	/		
lemon	///		

b Now represent the data in a waffle diagram. The waffle diagram will need a title and a colour-coded key.

Title: _____

Key

c   Is Alicia's prediction correct? [ ]   How do you know?

_____

_____

_____

d   What other conclusions can you draw from the data?

_____

_____

_____

_____

e   Now use the data in the completed chart in **a** to complete the pie chart. The pie chart will need a title and a colour-coded key.

Statistics and Probability

Statistics and Probability

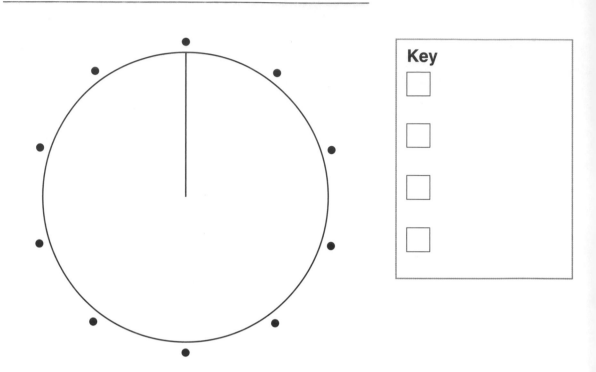

Key

3 Faisal counted the fruit in a bowl. He found the frequency of each type of fruit and converted this to a percentage. Then he used the data to draw a pie chart.

Use the information in the pie chart to answer the following questions.

a How many more apples are there in the bowl than oranges? Give your answer as a percentage. ☐ %

Percentage of fruit in a fruit bowl

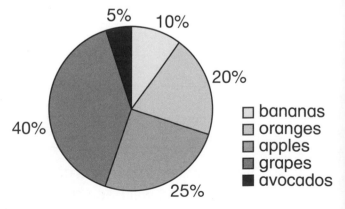

b How many fewer bananas are there in the bowl than grapes? Give your answer as a percentage. ☐ %

c Complete the missing word in the sentence below to describe the relationship between the number of bananas in the bowl and the number of avocados.

The number of bananas is ☐ the number of avocados.

Statistics and Probability

d Complete the missing word in the sentence below to describe the relationship between the number of apples in the bowl and the number of avocados.

The number of apples is ☐ times the number of avocados.

e If there were 100 pieces of fruit in the bowl, how many of the following fruit would there be:

bananas ☐   oranges ☐   apples ☐   grapes ☐

avocados ☐

f If there were 50 pieces of fruit in the bowl, how many of the following fruit would there be:

bananas ☐   oranges ☐   apples ☐   grapes ☐

avocados ☐

4 These are the distances in metres jumped by 20 competitors in a long jump event.

2·4	3·6	4·1	6·5	5·7	4·3	2·9	4·8	7·1	5·3
6·6	7·9	1·2	2·6	3·2	6·8	5·4	7·7	2·6	3·1

Millie watched the event and believes that more competitors jumped a distance of 3–3.9 metres than any other interval (1–1·9, 2–2·9, 3–3·9, 4–4·9, 5–5·9, 6–6·9, 7–7·9).

On a separate piece of paper, construct a tally chart and record the frequencies of each distance jumped.

Include and complete a column for 'percentage'. Use these values to draw a waffle diagram using Resource sheet 14. Remember to give it a title and a key.

What conclusions can you draw from the data?

_____

_____

_____

Date: _____   ☺ ☺ ☹

Lesson 4: **Mode, median, mean and range**

• Find and interpret the mode, median, mean and range of a data set

1 The following temperatures were recorded in a school playground during May:

> 20 °C, 23 °C, 23 °C, 24 °C, 19 °C, 25 °C, 22 °C, 25 °C,
> 25 °C, 26 °C, 26 °C, 27 °C, 27 °C, 27 °C, 25 °C, 25 °C,
> 24 °C, 25 °C, 26 °C, 25 °C, 24 °C, 25 °C, 23 °C, 23 °C,
> 22 °C, 22 °C, 22 °C, 21 °C, 20 °C, 22 °C, 23 °C

a Construct a frequency table for each temperature.

Temperature	Frequency

b Find the temperature that is the mode of this data set. [   ] °C

c Find the range of the data set. [   ]

Statistics and Probability

Statistics and Probability

2 Find the mean, median, mode and range of each data set. Show your working out below.

a

Football scores
4, 2, 1, 2, 3, 5, 4

Mean [ ]   Median [ ]

Mode [ ]   Range [ ]

b

Hockey scores
6, 4, 8, 9, 5, 6, 4, 6

Mean [ ]   Median [ ]

Mode [ ]   Range [ ]

c

Baseball scores
4, 7, 6, 8, 3, 5, 7, 8, 6, 8

Mean [ ]   Median [ ]

Mode [ ]   Range [ ]

d

Video game scores
77, 66, 49, 58, 75, 66, 56, 58

Mean [ ]   Median [ ]

Mode(s) [ ]   Range [ ]

Statistics and Probability

3 Find the mean of each set of sales made at a sports shop. Show your working out below.

$7.00	$12.00	$15.00	$33.00	$50.00
$11.00	$12.00	$20.00	$29.00	$49.00
$14.00	$9.00	$23.00	$24.00	$51.00
$12.00	$14.00	$21.00	$26.00	$47.00
$16.00	$9.00	$19.00	$32.00	$48.00
Mean	$10.00	$17.00	$41.00	$49.00
	Mean	$25.00	$25.00	$52.00
		Mean	$30.00	$50.00
			Mean	$54.00
				Mean

4 The manager of a shoe shop wants to know how well a new brand of trainer is selling and which sizes are most popular.

She looks at a record of the number of each size sold to predict sales for the next week. This will ensure the shop has enough of each size in stock.

Here is a list of all sizes sold:

6, 6, 7, 7, 7, 8, 8, 8, 8, 8, 9, 9, 9, 10, 10, 11, 11, 11, 12, 12

a Calculate the mean, mode, median and range of the data. Show your working out.

Mean [ ]

Mode [ ]

Median [ ]

Range [ ]

[   ]

b Which of the three averages – mode, median and mean – is most useful to the manager? [ ] Why? Convince the manager.

_____

_____

_____

5 Solve the following problems.

a The mean of four numbers is 6 and the mode is 5. What could the four numbers be?

[   ]

b There are six numbers with a mean of 48. After removing one of the numbers, 33, what is the mean of the remaining numbers?

[   ]

c The smallest number in a set of seven numbers is 10. The range is 9. The mode is 12 and 14. The mean is 14. What are the six numbers?

[   ]

d The mode of four numbers is 7. If the median is 7·5 and the mean is 9, what are the four numbers?

[   ]

Date: _____

Lesson 1: **Frequency diagrams**

Statistics and Probability

- Plan and conduct an investigation for a set of related statistical questions
- Predict the answer to a statistical question
- Represent data in a frequency diagram

You will need
- ruler
- coloured pencil

1 Class 6C investigated the statistical question: 'What is the longest distance a learner in our class can flick a counter with another counter?'
Ria says, 'My prediction is 90 cm.'

Distance (cm)	Frequency
20–40	6
40–60	3
60–80	5
80–100	9
100–120	7

a   Does the data help to answer the statistical question? ☐ How?

b   Is Ria's prediction correct? ☐ How do you know?

2 Akihiro wants to know the average age of the people attending his dance class. He predicts that there will be more people aged over 36 than under 36. The table lists the ages of the people attending the class.

a   Use the data to construct a frequency diagram.

Age (years)	Frequency
16–20	5
20–24	2
24–28	7
28–32	5
32–36	4
36–40	3
40–44	6
44–48	4
48–52	7

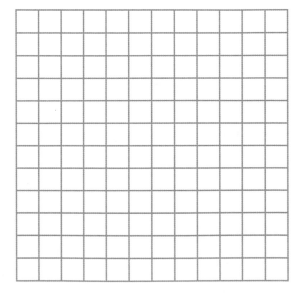

Statistics and Probability

b Is Akihiro's prediction correct? ☐ How do you know?

_____

4 **c** Write two conclusions you can draw from this data.

i _____

_____

ii _____

_____

3 Jessica records the number of hours of sunshine each day for 30 days.

She says, 'I predict that most days will have around 5 hours of sunshine.'

Jessica lists the number of hours of sunshine in a table.

4·5	5	3·5	7	1·5	5·5	4	4	2·5	6
4·5	7·5	5	3·5	0·5	8·5	6	8	7	4·5
2·5	2·5	3·5	5	4	4	5·5	3	3·5	4

a Complete the frequency table using these results. Decide on the most suitable intervals for the data.

b Use the data to draw a frequency diagram.

Sunshine (hours)	Frequency

c Is Jessica's prediction correct? ☐ How do you know?

_____

Statistics and Probability

④ **d** Write two conclusions you can draw from this data.

i _____

_____

ii _____

_____

4 Conduct an investigation similar to **1**.

③ **a** Write your statistical question here: _____

_____

b What is your prediction? _____

③ **c** Gather your data and present it here in a frequency diagram.

④ **d** What conclusions can you draw from the data?

_____

_____

_____

④ **e** Does the data agree with your prediction? [ ] How do you know?

_____

_____

Date: _____

☺ 😐 ☹

Lesson 2: **Line graphs**

- Plan and conduct an investigation for a set of related statistical questions
- Predict the answer to a statistical question
- Represent data in a line graph

You will need
- ruler

1 The graph shows the altitude of an aircraft from 0 to 10 minutes.

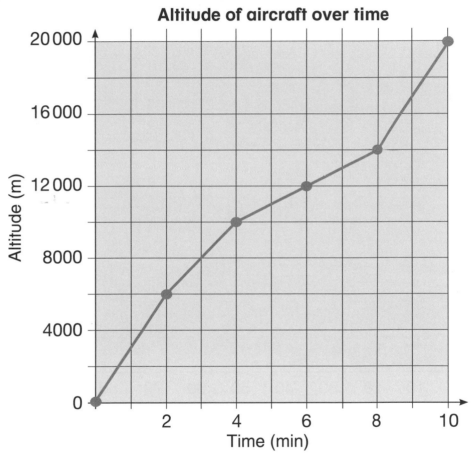

Altitude of aircraft over time

a What was the altitude of the aircraft in metres:

   i at 2 minutes? _____   **ii** at 4 minutes? _____

   iii at 8 minutes? _____   **iv** at 10 minutes? _____

b How much higher was the aircraft at 8 min than at 2 min? _____

c What was the approximate altitude of the aircraft at 9 min? _____

Statistics and Probability

2 The table shows the altitude of a hot-air balloon to the nearest 1000 metres from 0 to 90 min.

Altitude (m)	0	1000	2000	4000	6000	8000	12 000	16 000	18 000	20 000
Time (min)	0	10	20	30	40	50	60	70	80	90

a Plot a line graph on the grid to represent the data.

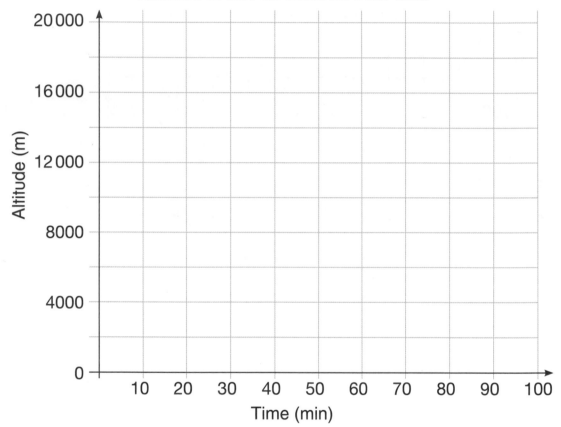

Altitude of hot-air balloon over time

b Between which times did the altitude of the balloon show the greatest change?

c Use the graph to find the appoximate height of the balloon at these times:

i 15 min

ii 55 min

iii 25 min

iv 85 min

3 Conduct an investigation in which you measure something that changes over time, such as the number of learners in a school corridor or the number of birds outside the window.

a Write your statistical question here: _____

_____

b What is your prediction? How do you think the data will change over time?

_____

c Gather your data and present it here in a line graph.

d Write two conclusions you can draw from the data.

i _____

_____

ii _____

_____

e Does the data agree with your prediction? [   ] How do you know?

_____

_____

Date: _____

Statistics and Probability

217

Statistics and Probability

Lesson 3: **Scatter graphs**

- Plan and conduct an investigation for a set of related statistical questions
- Predict the answer to a statistical question
- Represent data in a scatter graph
- Know how to draw a line of best fit

You will need
- ruler
- squared paper

1 Draw a line of best fit on each scatter graph. The idea is to draw the line so that there are an equal number of points above it as there are below it.

a

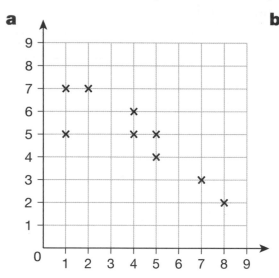

b

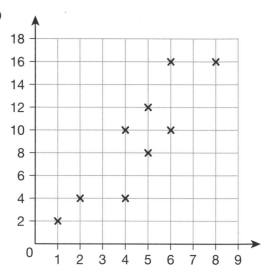

2 Lewis owns a computer shop. He collects data that records the number of customers visiting his shop and the number of computers he sells.

a Some data has not been included in the graph. Copy the graph on the right without the line of best fit, extend the axes and plot the following points: (35,8), (50, 10), (75, 14) and (85, 16).

Sales figures against customer numbers

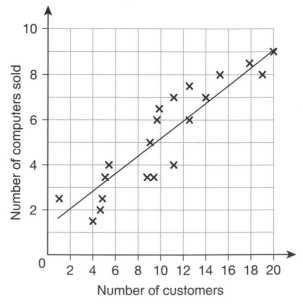

Statistics and Probability

b Is there a connection between the two sets of data? How do you know?

_____

_____

c Describe the relationship between the two sets of data.

_____

_____

d When there are 80 customers in the shop, how many computers should the shop expect to sell? [   ]

e Write two questions that can be answered using the graph.

i Question: _____

_____

Answer: _____

_____

ii Question: _____

_____

Answer: _____

_____

3 14 learners record the amount of time they spend completing homework and watching TV.

Time spent completing homework (min)	Time spent watching TV (min)	Time spent completing homework (min)	Time spent watching TV (min)
45	30	35	40
50	25	40	25
35	35	20	50
15	55	55	30
25	50	45	40
50	20	15	45
60	20	25	45

Statistics and Probability

a Plot a scatter graph to see if the two are related.

b What is your prediction? Explain your answer.

_____

_____

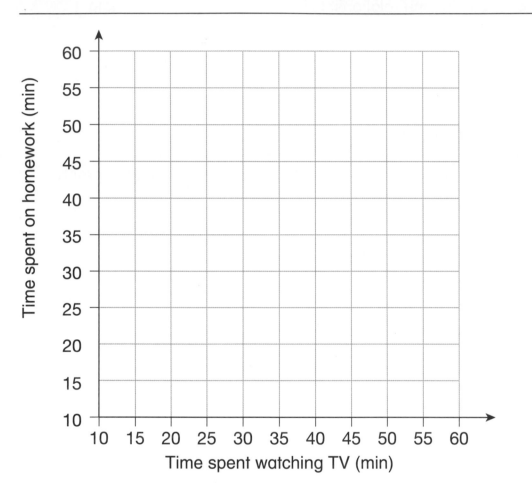

c Write two conclusions you can draw from the data.

i _____

_____

ii _____

_____

d Does the graph support your prediction? [ ] How do you know?

_____

_____

Date: _____

Statistics and Probability

Lesson 4: **Dot plots**

- Plan and conduct an investigation for a set of related statistical questions
- Predict the answer to a statistical question
- Represent data in a dot plot

1 A count was made of the number of birds on the branches of a tree.

The data collected was used to draw a dot plot.

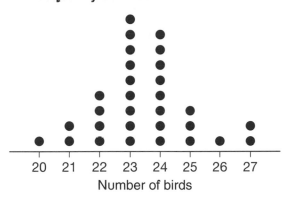

Frequency of birds on a tree branch

Number of birds

a How many branches had 22 birds?

b Which number of birds was only found on 3 branches?

c How many branches had over 23 birds?

d How many branches had fewer than 25 birds?

2 A count was made of the number of movies each family watched over a month. The data collected was used to draw a dot plot.

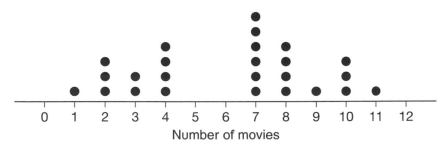

Number of movies watched last month

Number of movies

Statistics and Probability

a Which number of movies has the highest frequency? ☐

The lowest frequency? ☐

b What is the range of the data? ☐

c Does the data show any peaks, clusters or gaps? If so, where?

_____

_____

d Write two conclusions you can draw from this data.

_____

_____

3 The manager of a biscuit company checks that her machines are packaging very close to 50 biscuits per box.

She counts the contents of 30 packets.

49	46	50	49	49	45	44	50	46	50
44	47	45	46	48	44	45	46	45	49
49	50	46	44	43	46	44	45	45	46

a Complete the tally chart for this data.

Number of biscuits	Tally	Frequency

b Draw a dot plot to display the data.

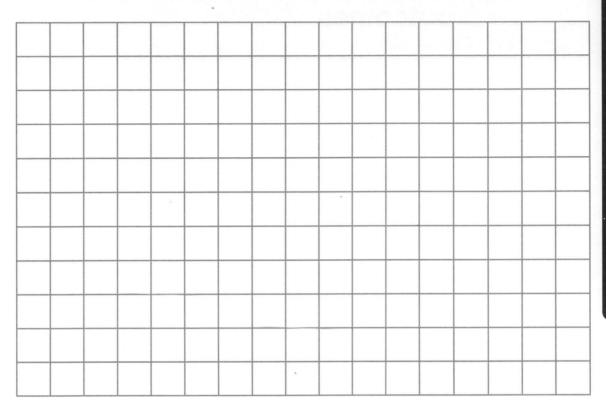

c Which number(s) of biscuits has the highest frequency?

d Which number(s) of biscuits has the lowest frequency?

e What is the range of the data?

f Does the data show any peaks, clusters or gaps? If so, where?

_____

_____

_____

g Should the manager be concerned? If so, why?

_____

Date: _____

Statistics and Probability

Lesson 1: **Describing and comparing outcomes**

• Use the language of proportion and probability to describe and compare the probability of different outcomes

1 For each event, tick the likelihood of it happening.

Event	Probability		
	Impossible	**Even chance**	**Certain**
spin an 11 with a 1–10 spinner			
spin a number less than 11 with a 1–10 spinner			
spin an even number with a 1–10 spinner			
spin a number less than 6 with a 1–10 spinner			

2 Give your answers as a proportion and a percentage.

What is the probability of picking a:

a square? ☐ in ☐ or ☐ %

b circle? ☐ in ☐ or ☐ %

c triangle? ☐ in ☐ or ☐ %

3 a Complete the table.

Event	Outcome	Probability
A	spinning a 2	1 in 10 (10%)
B	spinning a 1	
C	spinning a 4	
D	spinning a 3	

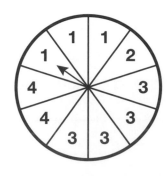

Event	Outcome	Probability
E	spinning a number greater than 2	
F	spinning a number less than 4	
G	not spinning a 3	

b Compare the probabilities of the events using the symbols, >, < or =.

i A ☐ B   **ii** E ☐ C   **iii** A ☐ F

iv G ☐ E   **v** F ☐ D   **vi** E ☐ B

4 Label the spinner with the letters: 'A', 'B', 'C' and 'D' so that the spinner follows these probability rules:

There is 1 in 8 chance of spinning a C.

There is 25% chance of spinning an A.

There is 1 in 8 chance of spinning a D.

There is 50% chance of spinning a B.

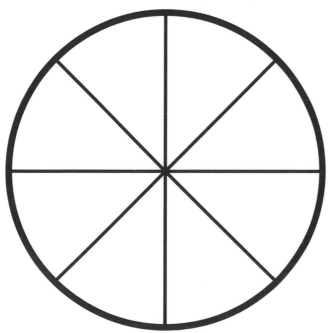

Date: _____

Statistics and Probability

Statistics and Probability

Lesson 2: **Independent and mutually exclusive events**

* Identify when two events are mutually exclusive or independent

1 Tammy rolls a 1 to 6 dice and it lands on a '4'. Write an event that:

a **cannot** happen at the same time.

_____

b **can** happen at the same time.

_____

 2 Complete the table by writing 'M' for a mutually exclusive pair of events
6 and 'I' for events that are independent. Give a reason for your choice.

Event and outcome	Mutually exclusive (M) or Independent (I)	Reason why
A 1 to 6 dice is rolled. The number is odd and a multiple of 2.		
A 4-letter spinner ABCD is spun and a coin is flipped. The letter is C and the coin is 'tails'.		
Winning a cricket match and drawing the same match.		
A 50-page book is opened at a random page. The page is 49 or 50.		
Two coins are flipped. The first coin is 'heads' and the second coin is 'heads'.		
On a board numbered 1 to 25, the first dart hits 20 and the second dart hits 20.		

3 Write a pair of events that have independent outcomes involving flipping a coin and rolling a 1 to 6 dice.

_____

_____

_____

4 Write a pair of events that have mutually exclusive outcomes involving picking a card at random from a set of 26 cards of the alphabet.

_____

_____

_____

5 Each pair of events is mutually exclusive. What is the probability of each event?

a Rolling a 1 to 6 dice and getting an odd number. ☐

Rolling a 1 to 6 dice and getting an even number. ☐

b Flipping a coin and getting 'heads'. ☐

Flipping a coin and getting 'tails'. ☐

c Spinning an ABCD spinner and getting a 'D'. ☐

Spinning an ABCD spinner and **not** getting a 'D'. ☐

i Find the sum of each pair of probabilities. What do you notice?

_____

_____

ii How would you explain this?

_____

_____

Statistics and Probability

Date: _____

227

Lesson 3: **Event probability and the number of trials (1)**

- Predict and describe the frequency of outcomes using the language of probability
- Perform probability experiments using small and large numbers of trials

1 Oba spins the spinner on the right. What is the probability that she will spin a:

a '1': ☐ in ☐ or ☐ %   **b** '2': ☐ in ☐ or ☐ %

c '3': ☐ in ☐ or ☐ %

d '1' or a '3': ☐ in ☐ or ☐ %

e '2' or a '3': ☐ in ☐ or ☐ %

2 Billy rolls a 1 to 8 dice.

He rolls the dice 60 times and records the number each time.

a How many rolls would you expect Billy to record a number less than 3?

b How many rolls would you expect Billy to record a number greater than 2?

3 Lucy spins the spinner shown 200 times and records the colour each time.

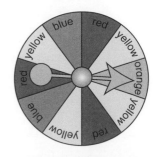

Statistics and Probability

Statistics and Probability

a How many spins would you expect Lucy to record 'red'?

b How many spins would you expect Lucy to record 'yellow'?

4 Arthur spins the spinner shown 80 times.

a Complete the table to show Arthur's predictions.

Outcome	Predicted probability (over 80 spins)	Predicted frequency of spins (over 80 spins)
spinning an 'A'	%	
spinning a 'B'	%	
spinning a 'C'	%	

Arthur completes the 80 spins. The table below shows the results.

Outcome	Experimental frequency of spins (over 80 spins)
spinning an 'A'	10
spinning a 'B'	27
spinning a 'C'	43

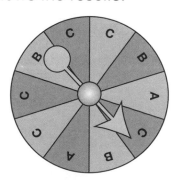

b The results are different from the predicted frequencies. Why do you think this is?

_____

_____

Date: _____

Statistics and Probability

Lesson 4: **Event probability and the number of trials (2)**

- Predict and describe the frequency of outcomes using the language of probability
- Perform probability experiments using small and large numbers of trials

1 Lucas spun a 4-colour spinner 100 times and recorded the outcomes. Write the missing numbers in the table.

Colour of sector	Predicted probability	Frequency	Experimental probability
blue		30	
yellow	$\frac{1}{4}$ (25%)	25	
red			$\frac{1}{5}$ (20%)
orange		25	

2 a Tammy rolls a 1 to 8 dice, 80 times.

i What is the predicted probability of rolling a number less than 3?

[ ] in [ ] ([ ] %)

ii What frequency would you expect Tammy to roll a number less than 3 over 80 rolls? [ ]

iii What is the predicted probability of rolling a '7'? [ ] in [ ]

iv What frequency would you expect Tammy to roll a '7' over 80 rolls? [ ]

v What is the predicted probability of rolling an even number?

[ ] in [ ] ([ ] %)

vi What frequency would you expect Tammy to roll an even number over 80 rolls? [ ]

Statistics and Probability

b Tammy conducts the experiment and records the outcomes.

What do the experimental results show? Compare them to the predicted probabilities.

i Probability of rolling a number less than 3.

_____

_____

_____

_____

Dice roll (outcome)	Frequency
1	9
2	6
3	10
4	13
5	9
6	11
7	8
8	14

ii Probability of rolling a '7'.

_____

_____

_____

iii Probability of rolling an even number.

_____

_____

_____

 3 Harry used a spinner in a probability experiment to test the prediction: One out of every five spins will be the letter 'D'.

He completed the experiment and recorded the outcomes of 100 spins.

Spin (outcome)	Frequency
A	22
B	21
C	19
D	22
E	16

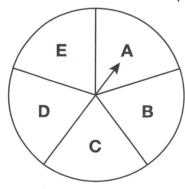

Statistics and Probability

a How well does the experimental frequency agree with the predicted frequency?

_____

_____

_____

_____

b Why do you think the predicted probability and experimental probability are not exactly the same?

_____

_____

_____

_____

4 Harry repeated the experiment for 1000 spins.

a What has changed about the results compared to 100 spins? How do you know?

_____

_____

Spin (outcome)	Frequency
A	178
B	212
C	205
D	190
E	215

_____

_____

_____

b What does this tell you about conducting an experiment with a larger number of trials compared to a smaller number of trials?

_____

_____

_____

_____

Date: _____